SCIENCE SKILLS

A Problem-solving
Activities Book

Alan Peacock

Routledge
Taylor & Francis Group

LONDON AND NEW YORK

To Githui, Githinji and Berluti

First Published 1986
by Macmillan Education Ltd

Published in 2013 by Routledge
2 Park Square, Milton Park, Abingdon, Oxon OX14 4RN
605 Third Avenue, New York, NY 10017

*Routledge is an imprint of the Taylor & Francis Group,
an informa business*

British Library Cataloguing in Publication Data

*A catalogue reference for this book is available from the British
Library*

ISBN 13: 978-0-415-09428-3 (pbk)

Publisher's Note
The publisher has gone to great lengths to ensure the quality of this
reprint but points out that some imperfections in the original
may be apparent

ACKNOWLEDGEMENTS

The author and publishers wish to acknowledge the
following photograph sources:
Alcan Jamaica Ltd p36 bottom; F J Aldridge pp11 bottom
right, 29 right, 32 top left, 32 bottom right, 43 a and c, 44 top
left, 44 top right, 53; British Airways p62; British Tourist
Authority pp11 centre right, 46 top right; Jim
Brownbill p36 top; J Allan Cash pp37 top, 46 top left, 61
right; Ron Chapman p44 bottom right; Format
Photographers/Maggie Murray p60; Philip Harris
Ltd p39; Irish Tourist Board p36 middle; Rod
Jennings p11 bottom left; Meteorological Office p43 b and
d; North of Scotland Hydro Electric Board p61
left; Valerie Randall p32 bottom left; Rolex Watch
Ltd p29 left; Times Newspapers Ltd p63; Vision
International/Anthea Sieveking p49 bottom right; WHO/P
Almsky p51; Zambia Information Services p11 top right.

The publishers have made every effort to trace the copyright
holders, but where they have failed to do so they will be
pleased to make the necessary arrangements at the first
opportunity.

Cover illustration by Ursula Sieger
Text illustrations by Ian Andrews

Contents

Introduction

Learning Science these days is much more about solving practical problems by doing investigations. At the same time, there is much less emphasis at the early stages on acquiring specialist knowledge than there used to be. This book sets out to help top primary and lower secondary pupils learn the skills of practical investigation, whilst minimising the teachers' need to collect, prepare or purchase materials, a job which takes up more time and resources than most teachers can reasonably spare at present.

The double-page 'self-teaching' units take one skill or topic at a time, and use only the kind of materials you could find in any classroom or home. Childrens' immediate surroundings are made use of wherever possible.

In Part One, the first page of each unit teaches a skill, using worked examples: the second page then tests and reinforces the skill, by asking pupils to 'Try Yourself' on carefully-structured examples. In this way, each unit builds on skills already acquired, so that the first half of the book develops the full range of Science Skills, from Asking Questions through to Improvising Apparatus.

In Part Two, examples are grouped round familiar topics such as Air, Water, Health, Habitats, Tools, Work, etc. In each of these units, the activities and questions use the skills learned in Part One. For instance, Unit 22 on Reproduction has six items, which between them cover the skills of Asking Questions, Observing, Recording, Interpreting, Analysing, Suggesting Explanations, and Designing Experiments. The book has been trialled and an independent assessment of pupils' achievement confirmed that the approach led to significant improvement in skill acquisition. To help with such assessment, two test papers have been included, in Part Three, each of which should take about one lesson for the average pupil.

The book can be used in several ways.

The obvious approach is to allow pupils to follow the 'Skills' units of Part One in sequence, at their own pace: the 'Topic' units can then be used as follow-up or extension material, either during or after this sequence.

To make it easy for teachers to 'dip in' and find examples which test a particular skill or topic, Grids are included in Part Four, the Teacher's section. In these Grids, every question is cross-referenced according to the skill(s) and topic(s) it covers: for example, question 7J about the water clock is designed to develop the skill of Analysing information, but it also relates to Improvising, and so 7J appears in more than one place on the Grid.

Another use for the material is as a 'Combined Science' foundation, either at primary or secondary level, as a preparation for subsequent splitting into separate subjects. Used like this, the book is excellent as a flexible and practical way of confirming and developing skills which might otherwise only be covered in an unsystematic or piecemeal fashion.

With the average 10-12 year old in mind, language has been kept simple, using short sentences and a minimum of terminology. A lot of use is made of charts, pictures and diagrams as the information-base for examples and questions. However, the units are not simply worksheets requiring single-word responses: questions ask pupils to tabulate, discuss, write, draw and choose other suitable ways of recording answers. Questions also reflect the multi-cultural society we live in, by using examples relating to other cultures and countries.

So the book is about learning those basic skills of scientific investigation which can later be applied in any branch of Science, as well as in other familiar problem-solving situations. And these skills are learned largely through the pupils' own self-directed, practical activity, which makes few demands on the teacher in terms of didactic teaching, explanation or provision of apparatus. Having worked through Science Skills therefore, neither pupil nor teacher should find the developing of Science Skills a threatening or difficult task.

PART ONE: SKILLS

UNIT 1

Asking questions

People have always wanted to fly. Before the first aeroplane, people tried many ways of getting up into the air. They were all trying to answer the question: 'How can men fly?'

The Wright brothers answered the question when they built the first aeroplane. They could not build it from a book because it had never been done before. They had to try things out and make discoveries.

When the Wright brothers were trying things out to answer their question, they were doing science. *You* are doing science when you are trying things out and making discoveries to answer a question. Your question will not be as big as the Wright brothers' but that does not matter.

Here is something for you to do.

A: Activity

Take a nylon comb and comb your hair about ten times. Turn a tap so the water runs out *slowly*. Hold the comb close to the stream of water.

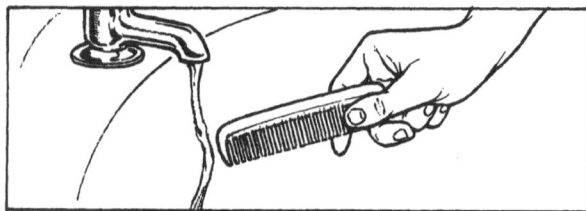

What happens? Did you ask any of these questions when you saw what happened?
Can I make the water bend the other way?
How far does it bend?
Does it always bend by the same amount?
Do all combs have the same effect?
Does it depend on the kind of hair?
What happens if the comb touches the water?

Answer these questions by trying things out. Try and make the water bend the other way. See how far you can make the water bend, and so on.

When you are trying things out in science you are doing experiments.

Sometimes you can answer a question just by observing or measuring. Sometimes a question will be very difficult and you will need more knowledge before you can answer it.

B: Activity

Make a paper dart like the one in the picture and fly it.

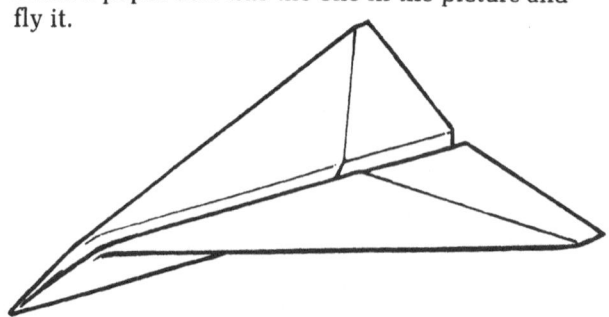

Does the dart fly well? You may have asked the question: 'How can I make it fly better?' This question may have led to other questions: Will it fly better if I fold the paper more carefully? Should the wings tilt forwards or backwards? Should the wings slope upwards or downwards? Are some kinds of paper better than others? Does the size matter?

Answer these questions by doing some experiments. The questions tell you what you need to do. Look at the first question above. To find the answer, make another dart by folding the paper very carefully. Fly it. Does it fly better than the first dart? You may have to fly the first dart again.

The *starting point* of most science is a *question*.
You were given some questions in the two activities on this page. Now you must learn to ask questions for yourself.

Try Yourself

C

Make a set of thumb and finger prints. To do this, put some ink on blotting paper. Press your finger on the ink and then on to clean paper.

Compare your prints with your friends' and classmates'.

Are any of the prints the same as yours?

What other questions come into your mind as you look at the prints?

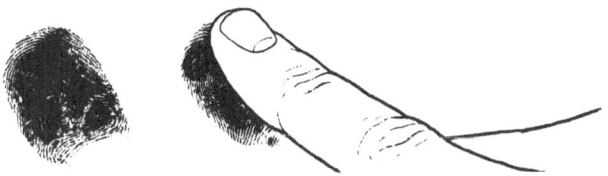

D

Some people can roll their tongues and some cannot.

Can you roll your tongue? Ask your classmates to try.

What questions can you ask to find out more about this? Here are two to start you off:

How many people in your class can do this?

Can you do it if you practise?

E

Fold your hands quickly without thinking about it. Which thumb is on top of the other? Is it the right-hand thumb or the left-hand?

How many right-hand thumbers are there in the class?

What other questions come into your mind?

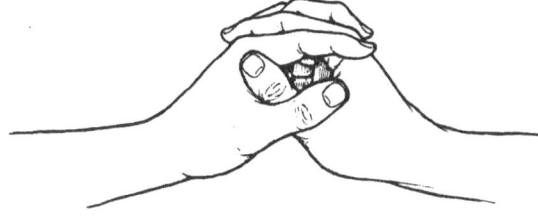

F

Find a partner and count the number of times that he or she breathes in 15 seconds. Then your partner should do 5 step-ups. (To do a step-up, find a *low* chair or box. Step up on to it and then step down on to the floor again *quickly*.) Count the number of times your partner breathes in 15 seconds after the step-ups.

What will happen to your partner's breathing rate if he or she does 10 step-ups?

What other questions come into your mind?

Your partner should measure your *pulse rate* when you are still. (Your pulse rate is the number of times your heart beats in one minute.) The diagram below shows you where to feel the pulse.

What questions come into your mind?

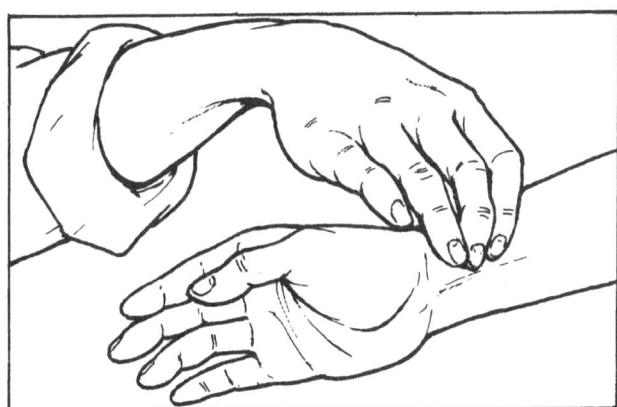

G

Go outside the classroom and make a collection of insects and tiny creatures. (Are you asking yourself, 'Where should I look?' 'How should I catch them?') Look under stones and in the grass. There are many other places to look too.

One way to catch the small animals is to hold a large sheet of paper under a tree branch. Then shake the branch hard. You can keep the small animals in matchboxes and jam jars.

When you have made your collection, study the insects for ten minutes or so. Write down *all* the questions that come into your mind about the creatures that you have collected.

UNIT 2
Observing

Observing means watching and noticing things around you. It is more than just looking because sometimes you look but do not see!

A: Activity

Close your eyes and try to describe your classroom.
How many windows are there?
How many desks or tables are there?
Which is the darkest or dustiest place?
  You must have 'seen' these things in your classroom many times but did you notice them?

This girl is observing sycamore leaves.

She is noticing several things so that she will recognise a sycamore tree next time she sees one. She is noticing the shape, size and colour of the leaves and the pattern of the veins.

B: Activity

Look at a 'BIC' pen like the one below.

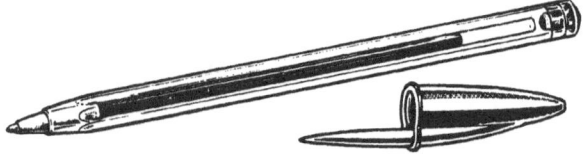

Observe it to find out what it is made of. You may notice that it has six different parts (not counting the ink). There are three different kinds of plastic and the tip is made of more than one kind of metal.
  Had you ever noticed these things before? If not, you have just observed your pen for the first time!

You can train yourself to be a better observer if you practise. Try this activity.

C: Activity

Observe the same 'BIC' pen again but this time look for signs that tell you if it is *old* or *new*. This time you may see things that you did not notice before. You may see the ink level, scratches on the side (and maybe tooth marks on the top!).

Sometimes you may be observing something *so* small that you need a *lens* or even a *microscope*.

D: Activity

Look at some sugar crystals through a lens. They will look something like this:

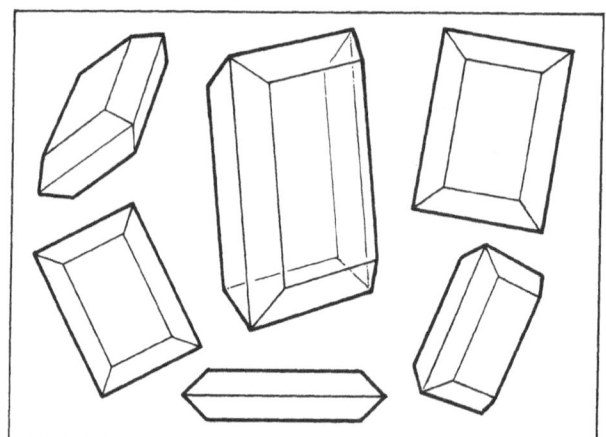

What do you notice about the *size* and *shape* of the crystals? They are not all the same size but they all have *edges* and *faces*. The opposite edges are *parallel*. Each crystal has 10 faces. Is the shape like a cube, or a diamond, or neither? Think of a word that describes the shape of a sugar crystal.

Observing is not simple. You need to have a *question* in your mind to observe well. Then when you look carefully at something you will see things that you had not seen before.
  In science you will use your other senses too. Listen, smell and touch *with a purpose*.

Try Yourself

E

How observant are you? Draw a picture of the front of your house. How many window panes are there in each window? Are there any drainpipes? What colour are the window frames?

F

Look out of your classroom window.
Look for:
1 the furthest object
2 the largest building
Listen for:
1 the sound that comes from furthest away
2 the direction it comes from

G

You will have seen zebras before but have you ever looked carefully at their stripes? Observe the zebra's stripes in the picture below.

Complete these sentences using the words given:
1 The broadest stripes are on its ....
2 The stripes are closest together on its ....
3 The stripes are crooked on its....
4 The stripes are horizontal on its ....
(face) (back) (rump) (legs) (tail)

H

Touch the window, the wall, the floor, the table and the door in your classroom. List them in order from *warmest* to *coldest*.

I

Look at some *salt crystals* with a lens.
1 How many *edges* are there on each crystal?
2 How many *sides* are there?
3 What *shape* are the crystals?

Make these observations on your way home or next time you go out.

J

How many lamp posts or telegraph poles do you pass on your way home from school?

K

Look at the reflection of yourself in a window. Hold up your right hand. Which hand is your reflection holding up?

L

How strong is the wind blowing? Write down five observations that give you the answer.

M

What colour is the bark on a tree? Observe the nearest tree. Write down all the colours you can see. Are you surprised?

N

Look closely at your school buildings. What materials were used to construct them?

O

Look at a lamp post in front of you on the other side of the road *but keep on walking on your side*. Observe the lamp post and the buildings or trees behind it as you walk past it. Does the lamp post seem to 'move' in front of the trees and buildings?
In which direction?

P

Next time a car or train passes you going very fast and sounding its horn, listen carefully to the horn. Does the sound change as the car or train goes by? How does it change?

UNIT 3

Looking for similarities and differences

If you look at an apple and an orange, you will see that they are alike in some ways. They are both round, and they have a skin and pips. But in other ways they are different. An orange has segments but an apple does not, and they are different colours.

We are *comparing* when we look for ways in which things are alike.

A: Example

How are these two kinds of leaves alike?
How are they different?
Make a list of how they are alike and how they are different. You will have to observe the drawings very carefully.

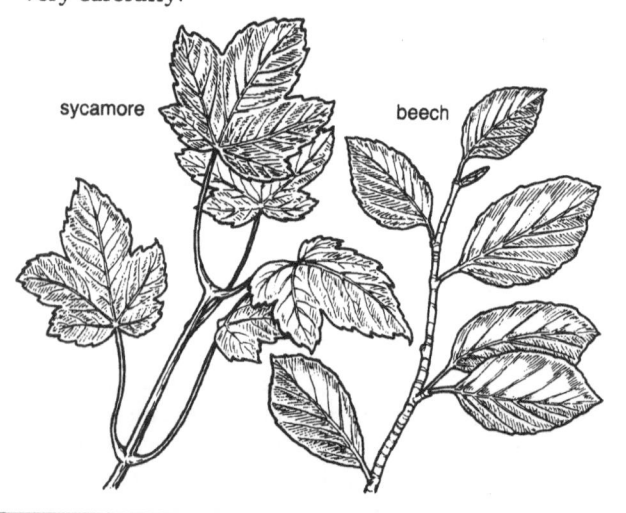

sycamore   beech

Which of these points did you observe in the example?
Similarities
The edges of the leaves are jagged.
Both kinds of leaves have veins.
Each leaf is attached to a central branch or twig.
The leaves get smaller towards the tip of the twig.

Differences
The sycamore leaf has five points; the beech leaf has only one.
The pattern of the veins is different.
The sycamore leaf is attached to the twig by a stalk; the beech leaf has no stalk.

B: Example

Here are four drawings of a child at different ages. Compare the size and shape of the different parts of the body at the different ages.

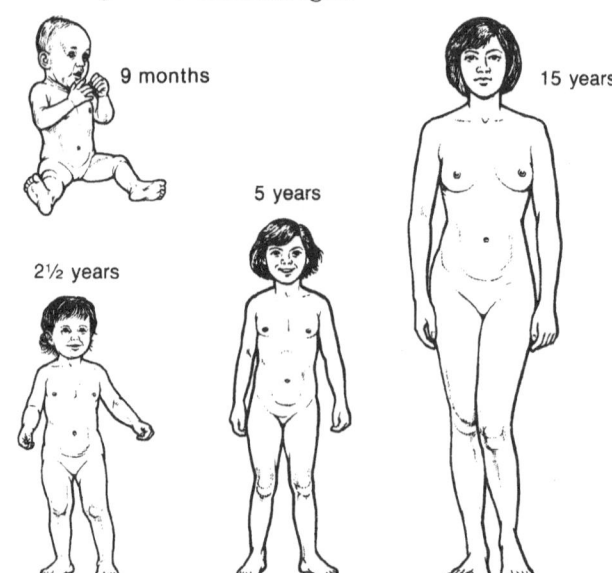

9 months   15 years

5 years

2½ years

The *proportions* of the body change as a child grows. This means that the head of a baby is big when compared with its body but the head of a fifteen year-old is not. The legs of a baby are short when compared with the rest of its body but the legs of a fifteen year-old are not.

Try Yourself

C

Compare these leaves from a gum tree with the beech leaves and sycamore leaves on the opposite page. Make lists to show:

1 the ways in which gum leaves are *similar to* sycamore and beech

2 the ways in which gum leaves are *different from* sycamore and beech

3 the ways in which gum leaves are *like beech* leaves but *unlike sycamore* leaves

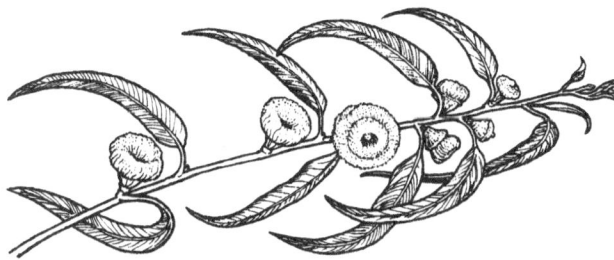

D

Here are pictures of two different kinds of

Look at the *materials* being used in the first picture. How are they different from the materials in the second picture?

How is the basket in the first picture being made? How is this different from the way the basket in the second picture was made?

You go to a shop to buy a basket. There are many different ones to choose from. What things would you compare before you bought one?

E

Look at these three pictures of road bridges. Each bridge supports the road in a different way. Compare the ways the road is supported.

UNIT 4

Classifying

Clive has emptied his piggy bank and is sorting the coins. He is putting them into piles of 1p pieces, 2p pieces, 5p pieces and 10p pieces. All the coins that are alike are put together.

At school, you are put into classes depending on how old you are. Most of the children in your class are the same age.

When we put people or things together because they are alike, we are *classifying*.

We can classify things in different ways.

A: Example

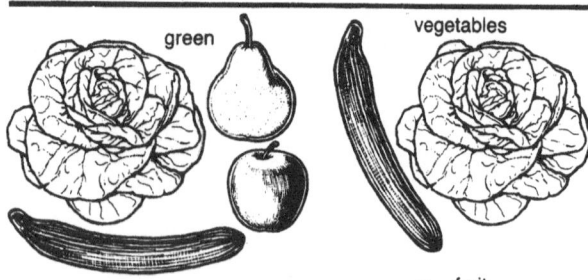

The groups on the left have been classified by *colour*. The groups on the right have been classified as *fruit* and *vegetables*. Which is the most *useful* classification?

You must be sure to put things into the right groups when you classify them.

B: Example

Here are three different groups of kitchen objects:

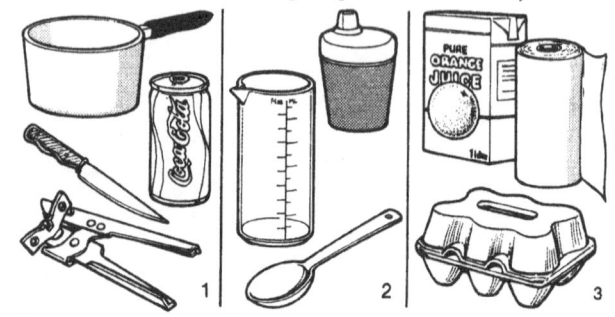

Group 1 contains objects made of *metal*.
Group 2 contains objects made of *plastic*.
Group 3 contains objects made of *paper*.
Which group would you put a yoghourt pot in?

We have to classify *living things* because there are so many of them. Here is one simple way of classifying them.

C: Example

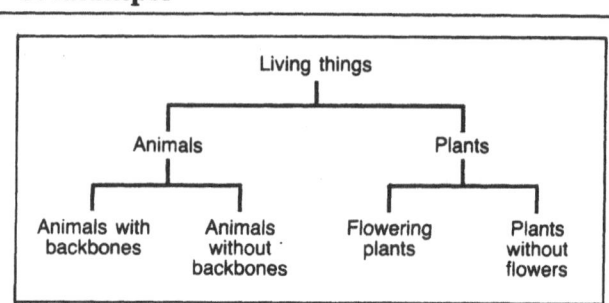

Each group can be divided up many more times. For example, animals with backbones can be classified as fish, birds, mammals and so on. Plants without flowers can be classified as seaweed, mushrooms and toadstools, ferns and so on.

When you classify, you decide what groups you want to put things into. Then you put all the things that are alike into the same group.

Try Yourself

D

Put these objects into two groups: things that let heat through easily (conductors) and things that do not (insulators).

glass bowl

tea cosy

steel spoon

thermos flask

aluminium saucepan

wooden bread board

copper kettle

E

Lee tried to classify some things at home as either *cutters* or *levers*. But he found that some things were both. These were his groups:

Cutters	Cutters and levers
knife	nail clippers
razorblade	scissors
potato peeler	tin opener

Levers
tap
door handle
tongs

Which of the three groups would you put these into: bottle opener, key, corkscrew?

F

Mike and Julie have been given a long shopping list by their mother. They decide to make a new list. They put all the things they will buy in the same shop together, like this:

Butcher's shop
sausages
lamb chops
bones for the dog
bacon

Finish their new list for them. Here are the other shops they will go to:

Greengrocer's shop
Do-it-yourself shop
Post office
Garden centre

bacon
oranges
wood glue
nuts
lamb chops
wire
cabbage
stamps
beans
sausages
potatoes
two pot plants
air mail letters
hooks
pineapple
bones for the dog
screws
lettuce seeds
onions

UNIT 5

Recording

There are many different ways of *recording* what you do and observe.

Suppose that you are doing a project about animals. If you want to record what you do, you will probably *write* about it. You may keep a *diary* like the one below. You might even use some *photographs*.

Writing a diary

You might *count* some things in your project. For example, you might count how often you observe a particular bird. You might count how many different animals you see in one place. The best way to record the *numbers* you get is in a *table* or a *chart*.

Making charts

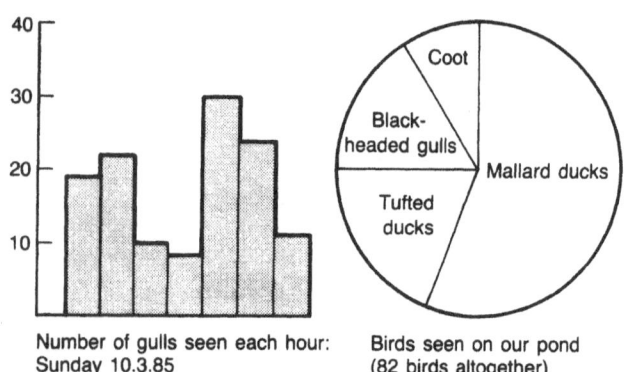

Number of gulls seen each hour: Sunday 10.3.85

Birds seen on our pond (82 birds altogether)

You might want to record the *sounds* that different animals make. You could write about them, but it would be much better to use a *tape recorder*.

You can make *drawings* or take *photographs* to record what the animals look like. You can make a drawing only if the animal stays still for long enough. You may have to take a photograph if the animal is frightened and moves quickly, or simply make some *notes*.

Making drawings

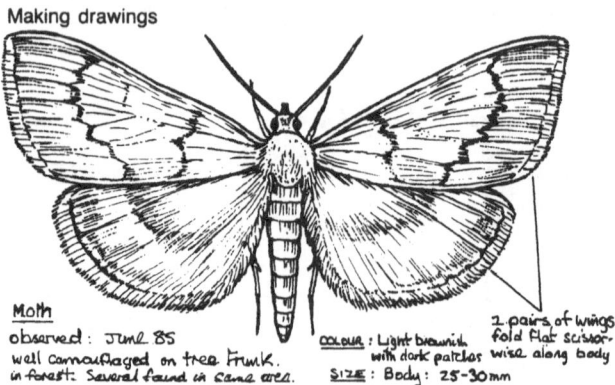

Moth

observed: June 85
well camouflaged on tree trunk in forest. Several found in same area.

COLOUR: Light brownish with dark patches
SIZE: Body: 25-30mm

2 pairs of wings fold flat scissorwise along body

Your project may involve a visit to a farm or zoo. You could record the visit in *photographs* and a *story*. You could make a *model*, or a *display* of the things you brought back. You may see something very complicated: you could draw a *diagram* to make it easier to understand and you could write some notes.

Making a display

You have to *choose the best way* to record your observations. If you are not sure of the best way, discuss it with your teacher and friends.

Try Yourself

A

Gary observed a blackbird's nest all morning. The things that he observed are listed below on the left. Some possible ways of recording his observations are listed on the right.

Observations	Ways of recording
1 the nest and young 2 the bird's comings and goings 3 the behaviour of the young birds 4 the food brought by the parents	list drawing written description time chart

Match each observation with the best way of recording it.

B

Imagine that you are trying to describe the creature in this picture to a blind friend. Write down your detailed description.

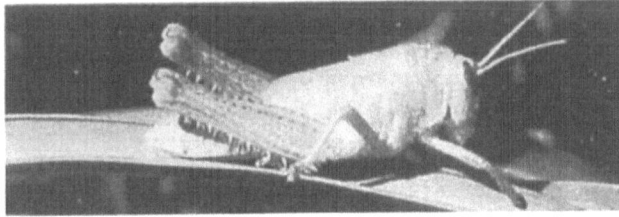

C

Study the hair colour of all the children in your class. Which is the most common colour? How many different colours are there? Show how you would record your observations.

D

The photograph below shows a home-made torch.

Describe how the torch was made by writing about it and drawing a diagram.

E

Owen tried to draw a goose and a hawk. As you can see, his drawings turned out nearly the same. Try to improve his drawings by using the information given below.

Geese are large, plump birds with long necks, short legs and round tipped beaks. They eat grain, grass shoots and seaweed. Their feet are webbed.

Hawks have sharp, hooked beaks, and toes with strong, curved talons. They have short necks and longish tails. They eat small, live animals such as mice.

F

Observe a simple machine such as a pencil sharpener, a zip or a tin opener. Choose the best way to record how it works.

G

How can you use *colour* to help you record things? Discuss this with your group.

H

Sandy wanted to find out how animals 'talk' to each other. She observed some dogs, birds, insects and sheep. She found that they could 'say' things like, 'Keep off!' 'Look out, danger!' and 'Look at me!'
How could Sandy record what the she observed?

I

Jamie and Stephen were doing Activity F on page 7 of this book. Jamie counted how many times Stephen breathed in 15 seconds when he was still. Then Stephen did 5 step-ups and Jamie counted the number of breaths that he took in 15 seconds again. Then Stephen did 10 step-ups and Jamie counted the number of breaths in 15 seconds once more.

Draw a table for them to record their results. Write in the headings.

UNIT 6

Interpreting

In Unit 5, you learned how to record information in writing, tables, graphs, diagrams and displays. *Interpreting* means 'reading' tables, graphs and diagrams to get as much information as possible from them.

A: Example

Here is some information about the temperature of different parts of a school at different times of the day.

	9 am	midday	3 pm
Classroom	15°C	18°C	20°C
Cloakroom	14°C	15°C	16°C
Hall	14°C	21°C	20°C

This table gives you nine pieces of information. The number in the dotted circle tells you that the temperature in the classroom at 9 a.m. was 15°C. What does the table tell you about the temperature in the hall at 3 p.m.? Look in the bottom right-hand box. You will see that it was 20°C.

B: Example

Look at the diagram below. What is it telling you?

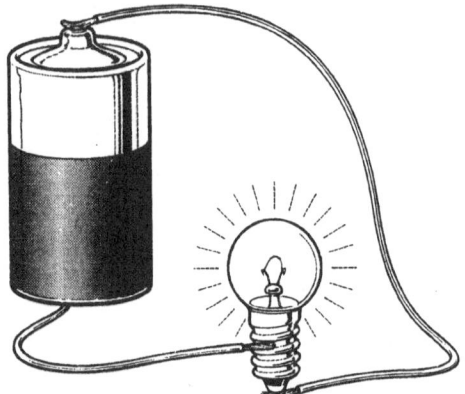

A *wire* is connected from the *top* of a *battery* to the *bottom* of a *bulb*. *Another* wire is connected from the *bottom* of the battery to the *side* of the bulb. The bulb *lights*.

C: Example

In this example, the graph tells you about the weights of two babies in the first year of their lives.

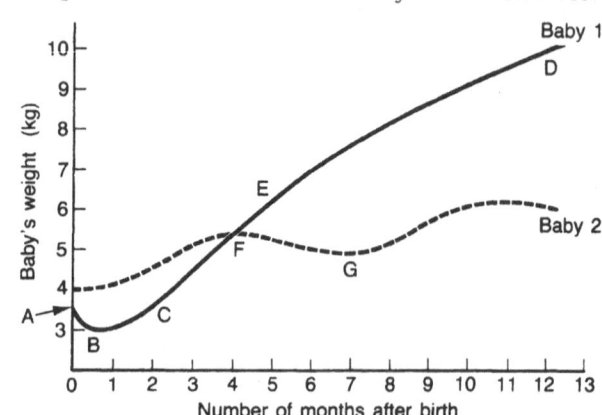

A graph like this tells you many things. For example, look at the solid line for Baby 1. It tells you the following things:
1. At birth, Baby 1 weighed 3.5 kg (point A on the graph).
2. After two weeks it had lost weight and was only 3.0 kg (point B on the graph).
3. The baby was two months old before it was the same weight as when it was born (point C on the graph).
4. It weighed 10 kg after one year (point D on the graph).
5. It grew fastest between the second and the fifth month (between points C and E).

You have seen that tables, graphs and diagrams contain a lot of information. But you have to learn to 'read' them to get all the information. This is called *interpreting*.

Now try interpreting for yourself.

Try Yourself

D

Look at the graph about babies' weights on the opposite page and answer the following questions.
1 What was Baby 2's weight at birth?
2 What was Baby 2's weight after 12 months?
3 What happened to Baby 2's weight between points F and G on the graph?
4 How old were the two babies when they weighed exactly the same?

E

One of these statements is false. Which one is it?
1 After one month, Baby 2 was heavier.
2 After two months, Baby 2 was growing faster than Baby 1.
3 After four months, Baby 2 began to lose weight.
4 After nine months, Baby 2 was about 3 kg lighter than Baby 1.

F

This table is printed on a tin of baby milk.
1 How many feeds should a 3 month-old baby have each day?
2 How many spoons of powdered milk should be put in one feed for a new-born baby?
3 What *volume* of milk should be given to a 4 month-old baby *each day*? (The answer should be in millilitres.)

G

The four diagrams below are found on the same tin of dried baby milk. Write the instructions which you think should go with them.

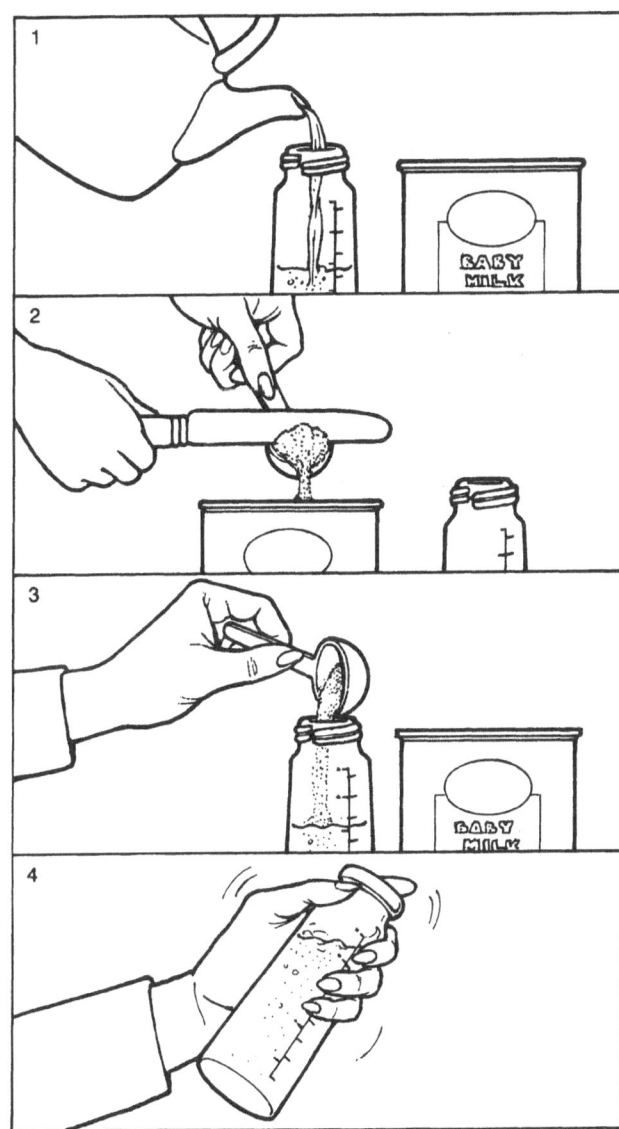

Weight of baby	Age of baby (approximately)	Number of spoons of milk powder	Volume of boiled water (ml)	Number of feeds per day
2.5 kg	0–2 weeks	3	85	7
3.5 kg	2–6 weeks	4	115	6
4.5 kg	2 months	5	140	6
5.5 kg	3 months	6	170	5
6.5 kg	4 months	7	200	5
8.0 kg	6 months and over	8	225	4

UNIT 7

Analysing

In Unit 6, you learned how to 'read' tables, graphs and diagrams to get all the information you could from them. In this unit, you will learn to *analyse* the information. This means that you will study the information to find *only* what you need to answer your questions.

A: Example

Look again at the table on page 16. You read and understood the information: different parts of a school are at different temperatures at different times of the day.

Suppose that some students have complained that their classroom gets hotter and hotter during the day. Look at the information in the table again. This time, you have a question: Does the classroom get hotter during the day?

(The classroom is 15°C at 9 a.m., 18°C at midday and 20°C at 3 p.m.)

B: Example

Here is a weather chart like the one you see on the T.V. There is a key at the side of it to tell you what the symbols mean.

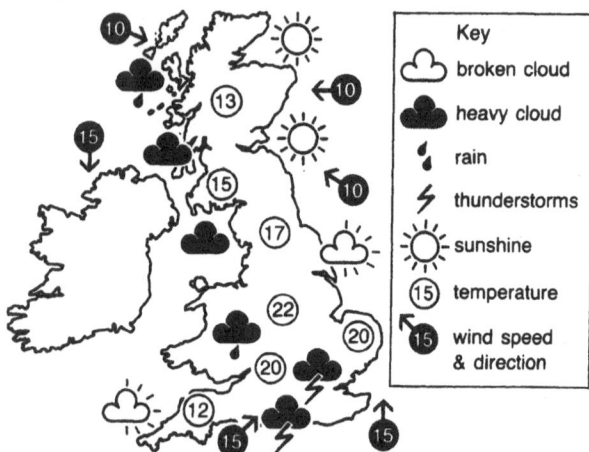

Key

☁ broken cloud

☁ heavy cloud

🌧 rain

⚡ thunderstorms

☀ sunshine

⑮ temperature

↖ wind speed & direction

Which coast is likely to have fine weather? Look for the sunshine symbols. Can you see a pattern along the east coast? What sort of weather can the west coast expect?

Try another example. Which direction is the wind blowing? Look at the wind arrows. They are not all pointing in the same direction but there is a pattern. The wind on this day is blowing from the sea towards the land.

Lots of the things we do are affected by the weather. Look at this next example.

C: Example

Tony likes fishing. He keeps a record of all his fishing trips in a notebook. From this, he hopes to find out which is the best time to fish. Here is a page from his notebook.

Date	Weather	Number of fish caught.		
		morning	afternoon	evening
May 5th	Sunny, dry cold	0	0	0
May 10th	Sunny, warm dry	0	0	2
May 11th	Cloudy, warm rain	2	0	3
May 16th	Cloudy, cold drizzle	0	1	1
May 18th	Cloudy, cold rain	0	1	0
May 23rd	Cloudy, dry warm	3	0	2

Tony's table tells you that he caught no fish on his first trip. He caught two fish in the evening on his second trip, and so on.

Tony wants to know *which kind of weather* is best for catching fish. Look at the page in his notebook. He caught most fish on cloudy, warm days. Perhaps this is the best weather for him to go fishing.

Suppose that Tony wants to know which time of day is best for catching fish. Look at the notebook again. He caught 8 fish in the evenings, 5 in the mornings, and only 2 in the afternoons. So evening was the best time and afternoon was the worst.

You can see that analysing is *more* than interpreting. It means that you are *studying the information* with a *question* in your mind.

Try Yourself

D

Douglas tried to make an *anemometer* for measuring the speed of the wind. He used the following things:

    plastic washing-up liquid bottle
    a cork
    3 thin knitting needles
    2 table tennis balls cut in
      half

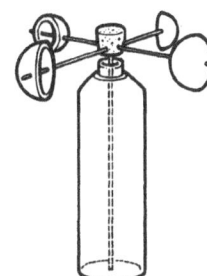

The diagram shows what it looked like when it was complete.
How is it supposed to work?

E

Douglas took it outside on a windy day and stood it on a wall. It didn't work. Can you think of 3 reasons why it would not work?

F

Discuss with your friends how you could fix the anemometer so that it worked properly.

G

Ken wanted to find out if eating sweets really does cause tooth decay. He found 20 pupils in his year that ate sweets every day; 20 that ate sweets 2 or 3 times a week and 20 that ate sweets once a week or less.

He asked each student how many fillings he or she had been given by the dentist. Then he worked out the average number of fillings for each group.

Sweets eaten	Every day	2 or 3 times a week	once a week or less
Average number of fillings for each group	6	3	2

From these results, do you think that eating sweets can cause tooth decay?

What other questions could Ken have asked the pupils?

H

One evening, Tony's dad said,
  'The weather forecast says sunny and warm for tomorrow. Shall we go fishing in the afternoon?'
  Tony looked in his notebook. What do you think he answered and why?

I

Emily made some gears by nailing bottle tops to a board like this.

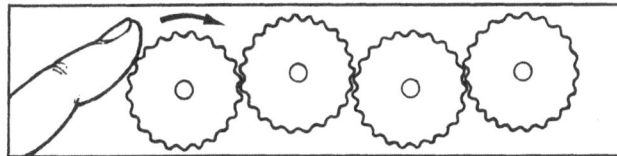

The arrow shows you that she turned the first bottletop clockwise. Which directions will the other bottletops turn in?

J

Martin tried to make a *water clock* from an empty plastic bottle. He punched a small hole in the bottom and filled the bottle with water. Then he made some marks down the side of the bottle with equal spaces between them. He decided to check his water clock against a real clock to see if it worked. The diagram shows what he found out.

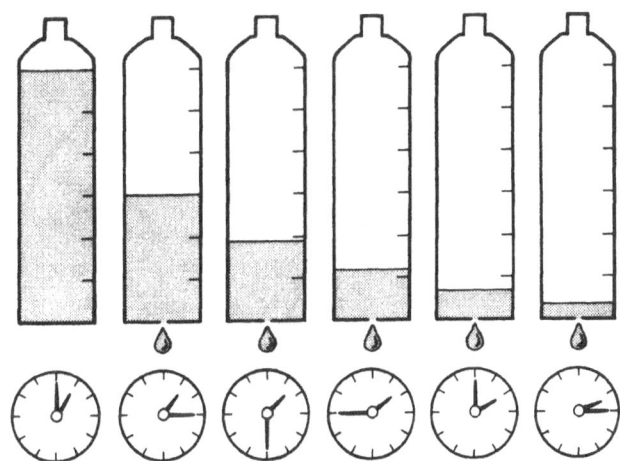

1 Is Martin's clock reliable? Give a reason for your answer.
2 At what time had half of the water run out?
(a) 1.10 (b) 1.15 (c) 1.25 (d) 1.30 (e) 1.40

UNIT 8

Concluding

In Unit 7, you learned how to analyse information. Did you notice that you learned *something new* from the information when you analysed it? For example, when you analysed the information in Tony's notebook, you found that the best time for him to go fishing was in the evening, in cloudy, warm weather. This is *new* information. You did not know this before you analysed the notebook.

The new thing that you learn is called a *conclusion*.

A: Example

Look again at the table of temperatures in a school on page 16. If you analyse the information, you can *draw these conclusions*:
1   The classroom gets hotter as the day passes. (You drew this conclusion without realising it!)
2   The cloakroom is always the coolest room.

B: Example

Look at the two pie charts below. They show the quantities of different kinds of fish in a river and a lake.

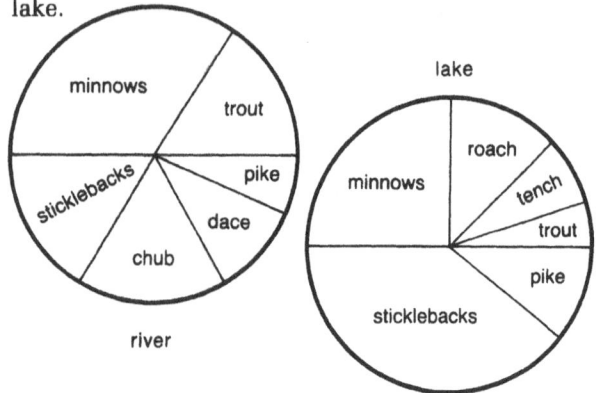

You can draw several conclusions from these two charts. For example:
1   Some kinds of fish live in both the river and the lake (pike, trout, minnows and sticklebacks).
2   Some kinds of fish live only in the river (chub and dace).

C: Example

Now try drawing conclusions from the diagram shown below. The diagram shows some things that live in and around a pond. The arrows show that some animals are eaten by others. For example, *flies* are eaten by birds, frogs and fish.

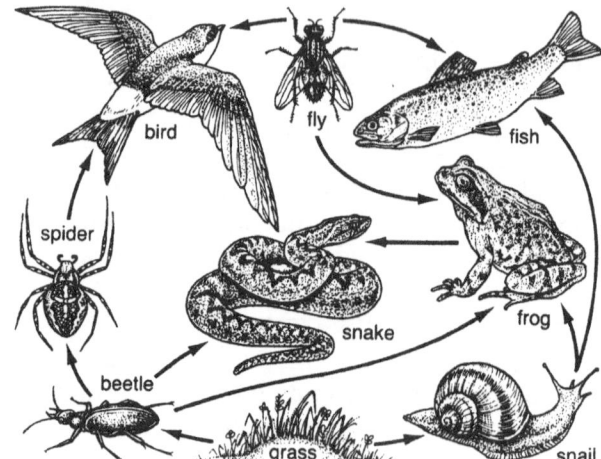

What would happen if all the frogs were killed? Analyse the diagram.

If all the frogs were killed, the snakes would not be able to feed on them. The snakes would have to eat more beetles. So, one conclusion you can draw is this: if all the frogs were killed, there would be fewer beetles.

Another conclusion you can draw is that snakes are at the *top of a food chain*. This means that snakes eat other creatures but nothing eats them.

In the last three units, you have learned how to make sense of information:
1   You have learned how to 'read' tables, graphs, charts and diagrams to get all the information you can (*interpreting*).
2   You have learned how to study information to get answers to questions (*analysing*).
3   You have learned to *draw conclusions* when analysing information.

Often when you look at information, you are *interpreting*, *analysing* and *drawing conclusions* all at the same time.

Try Yourself

D

Look at the diagram of the food chains on the opposite page again. Answer the questions below.

1 You saw that snakes are at the top of a food chain. Which other animals are at the top of a food chain?
2 What would happen if all the snails were removed from the pond?

E

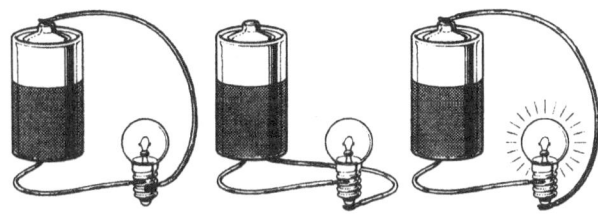

What conclusions can you draw from the diagrams above?

F

All of these are insects.

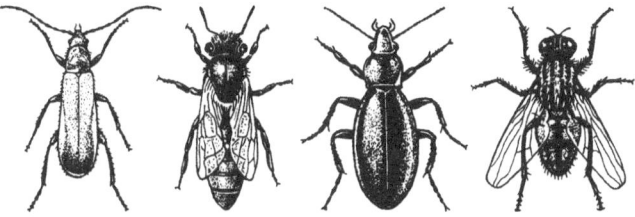

None of these are insects.

Which of these are insects?

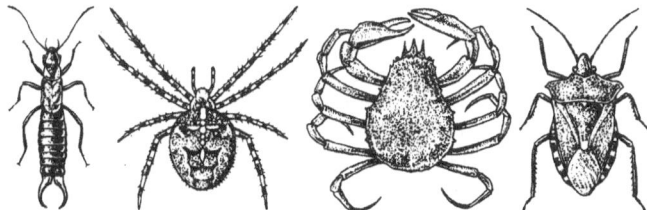

What conclusion do you draw about insects?

G

This table shows which textiles can be used to make clothes for different weather.

	Wind	Showers	Rain	Cold
Wool				X
Nylon	X	X		
Denim	X			
PVC	X	X	X	
Down	X			X

Choose clothes from the list on the right to match the types of weather listed on the left.

(1) very cold, windy and dry
(2) sunny, breezy and mild
(3) wet, windy and cold
(4) mild, showery
(5) warm and dry

(a) wool sweater and PVC jacket
(b) wool sweater and downy anorak
(c) nylon anorak
(d) denim jacket
(e) nothing extra

H

Margaret cut some strips of different kinds of paper. She put drops of different kinds of liquid on each one. She recorded her observations in diagrams which are shown below.

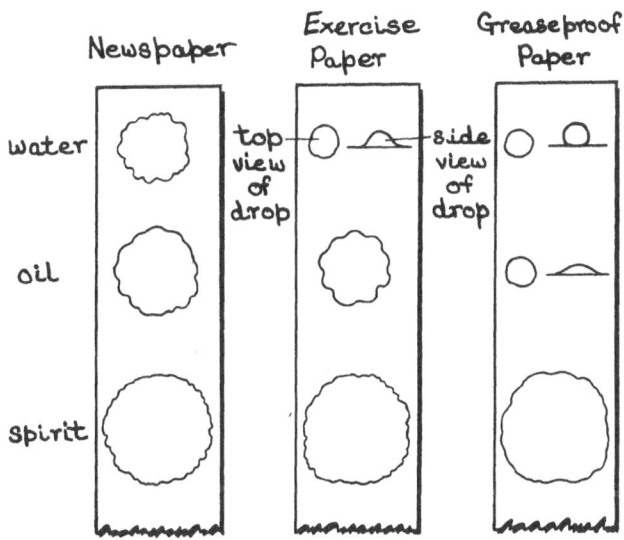

1 What conclusions could Margaret draw from the results of her experiment?
2 Why do you think greaseproof paper is called 'greaseproof'?

UNIT 9

Suggesting explanations

Trying to explain how or why things happen is an important part of science. Sometimes it is easy to explain why something happens.

A: Example

Food will cook more quickly in a pan *with* a lid than in a pan *without* a lid.

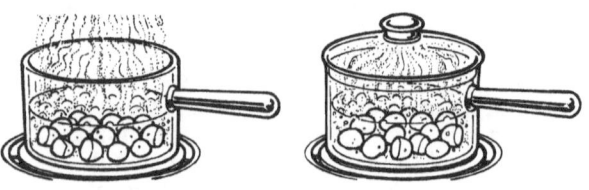

Why is this? The lid stops some of the heat from escaping.

Sometimes it is more difficult to find an explanation.

B: Example

Look at this row of plants on a window sill.

Why are some of the plants dying? Here are some *possible* explanations:
  They have too much sun
  They have not been watered enough
  They have been watered too much
  The soil is poor
  They have been attacked by pests
  They are in a draught
Which of these explanations is correct? To find out you might have to:
  Find out more about the plants
  Make some observations
  Do some tests
  Use what you already know about plants.

C: Example

Kerry mixed some water and cooking oil in a bottle and shook them up. After a while the oil and water separated. The oil settled on top of the water.

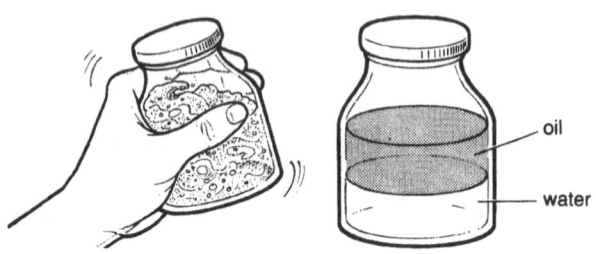

Kerry tried to explain why this happened. She knew from experiments she had done before that some liquids do not mix together. She also knew that lighter (less dense) liquids float on heavier (more dense) liquids. This is what she recorded in her notebook:

> I noticed that the cooking oil separated from the water and floated on top.
> Explanation: oil and water do not mix and oil is lighter than water so it floats on top.

D: Example

Barry picked some apples in the autumn and stored them in boxes. He wrapped each apple in newspaper as he packed the first few boxes. Then he got lazy and filled the last few boxes without wrapping the apples.
  Later in the winter, nearly all the apples in the last few boxes were rotten. But only a few of the apples that were wrapped in newspaper were bad. Can you suggest an explanation?
  Apples are made rotten by a *fungus*. The fungus can spread from apple to apple. This is why the apples in the last few boxes went bad. The newspaper around the apples in the first few boxes stopped the fungus from spreading.

Try Yourself

E

John and his friends were camping. They made some tea and poured it out into different cups. After a few minutes, John's tea was quite cool, but James's was still too hot to drink. Write down four possible reasons for this difference.

F

Sean was trying to find out how many nails would sink a bottletop floating on water. He dropped nails on to a bottletop until it sank. Then he repeated the test five times. These were his results:

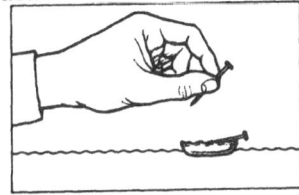

Test	1st	2nd	3rd	4th	5th
Number of nails	6	3	4	5	7

Give three possible reasons why the number of nails was not always the same.

G

Sue was trying to find out how quickly water drains through soil. She plugged some paper cones with cotton wool and put different kinds of soil into each one. She poured water on to the soil and timed how long it took for the water to start dripping through. These were the notes on her experiment:

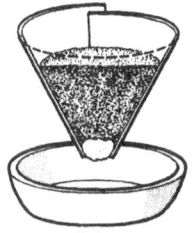  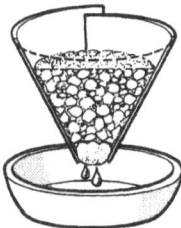

(a) no drips in 5 minutes   (b) begins to drip after 30 seconds   (c) water goes through immediately

Make a list of all the possible reasons why Sue got these different results.

H

If you put different objects on to a sloping surface, some will slide down but others will stay still. Which of the points listed below might explain why some objects slide but others do not?
1 the size of the object
2 its weight
3 its shape
4 the material it is made from
5 the smoothness of its surface
6 the smoothness of the slope
7 the height of the objects
8 the angle of the slope

I

Look at the diagrams below:

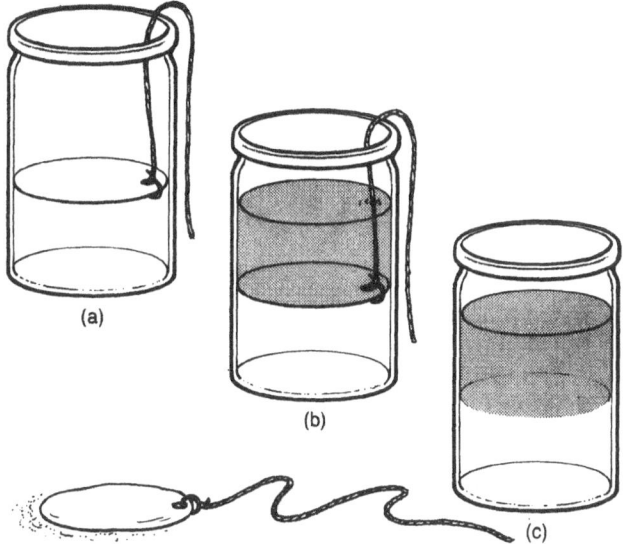

Diagram (a) shows a jam jar half full of cold water. A circle of paper is floating on the water. Diagram (b) shows the same jam jar with some hot, coloured water poured on top of the paper circle.
Diagram (c) shows the same jam jar after the paper circle has been carefully removed.
 Can you explain why the two layers of water do not mix?

J

In the first unit in this book, you looked for insects in different places. Think of where you found the insects. Discuss possible reasons why different insects like to live in different places.

UNIT 10

Predicting

In science, you will often ask questions that start like this:

'What will happen if . . .?'

These are *predicting* questions. When you try to answer a predicting question, you have to make a *sensible guess*. You use the knowledge you already have. A sensible guess is called a *prediction*.

A: Example

David floated an empty 'Coke' can and then started to put water into it using an egg cup. He observed the can each time he added one egg cupful of water.

David then asked himself, 'What will happen if I put another egg cupful of water in?'

You can see from the diagram that the can floats *deeper* as you put *more water* in. The can would probably float deeper still if one more egg cupful of water was added, like this:

B: Example

Suppose that somebody asks you, 'Will it rain today?' You might make a guess without thinking. But if you want to make a sensible guess, you will observe things that will help you. You will look at the *clouds*, the *wind direction*, the *weather chart* and the *barometer*.

Discuss the signs that *predict* rain with your group and your teacher.

C: Example

Brenda made a *pendulum* with some string and a stone. She found that when she changed the *length* of the string, the time of each swing changed. Here is her table:

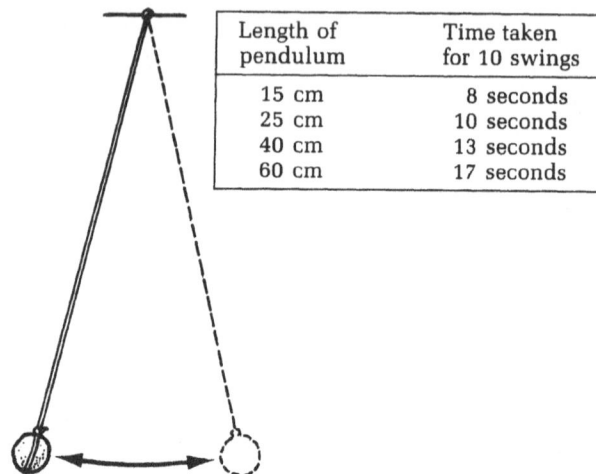

Length of pendulum	Time taken for 10 swings
15 cm	8 seconds
25 cm	10 seconds
40 cm	13 seconds
60 cm	17 seconds

Brenda wanted to know what would happen if she shortened the string to 10 cm. Her table shows that each time the string gets longer, the time for ten swings increases. She predicted that the time for ten swings of a 10 cm pendulum would be 7 seconds.

Brenda also wanted to know what would happen if she used a bigger stone. She realised that the tests she had done so far would not help her to predict the answer. She decided to do another experiment, using different sizes of stone.

Sometimes when we try to predict something, we are surprised by an unexpected result.

D: Example

Imagine that you fill one ice tray with cold water and another with hot water. Then you put them both into the freezer together. Which do you think will freeze first?

If you thought the cold water freezes first, you were wrong! The hot water freezes faster.

When you want to try and *predict* something, you look at *information that will help you*. Then you make a *sensible guess*.

Try Yourself

In the diagrams below, Jackie is doing some tests with stones and a jar of water.

Draw what happens next in the empty frame.

E

Look again at David's experiment with the Coke cans on the opposite page. How many egg cupfuls of water would be needed to make the can sink?

F

Look again at Sue's experiments with soil on page 23. What would happen if she poured more water through the same, wet cones of soil?

1 The second lot of water would not go through at all.
2 It would go through faster than the first lot of water.
3 It would go through more slowly than the first lot of water.
4 It would depend on other things and cannot be predicted.

G

Gill made two holes in a tin, and filled the tin with water. Diagram (a) shows what she observed. Then she made another hole in the tin, *between* the first two holes. (See diagram (b).)

How will the water run out now? Draw a diagram to show what you predict.

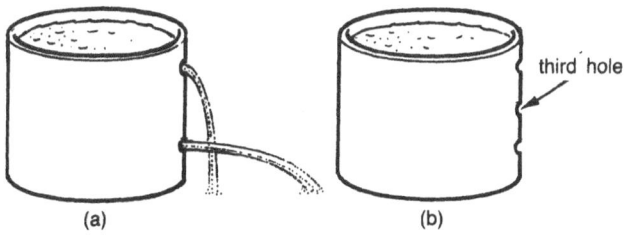

(a)   (b)

H

The table below shows how much petrol some different cars use, at different speeds.

Fill in the empty boxes in the table, using figures from this list: 16.2   23.8   29.7   36.6   43.5

Make of car	Engine size (cc)	Town Driving	Constant 56 mph	Constant 75 mph
Volkswagen Polo	900	32.8	47.9	35.3
Ford Escort	1300	31.0	☐	33.2
Vauxhall Cavalier	2000	23.7	39.8	☐
Rover 3.5	3500	☐	36.3	27.9

Figures taken from official Government figures, as quoted in *Motoring Which*, Vol. 20 (1981).

UNIT 11
Making tests fair

Sometimes you will have to do some tests before you can answer a question.

A: Example

Imagine that some of your plants are dying. Your question is, 'Why are some of my plants dying?'
- Perhaps you have not been watering them enough.
- Perhaps you have been watering them too much.
- Perhaps they are not getting enough sun.
- Perhaps they are getting too much sun.

To find out if you have not been watering them enough, you will have to give them more water. This is a test. If the plant recovers, you will have answered your question. Your plant was dying because you did not give it enough water.

B: Example

Here is a test that Tom did. He wanted to find out which kinds of beans sprout fastest. He put three different kinds of beans into containers, like this:

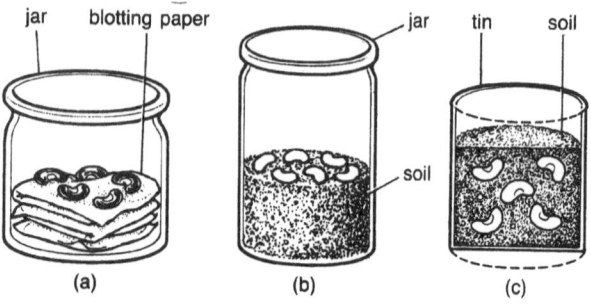

(a)   (b)   (c)

After a week, Tom noticed that the beans in jar (b) had sprouted first. He concluded that this kind of bean seed sprouted faster than the other kinds. But his test was *not a fair test*. Can you see why?

Tom's test was not fair because the beans were not grown in the same way. For example:

The seeds in (a) were grown on wet blotting paper, but the others were grown in soil.

The seeds in (c) were under the soil, but the seeds in (b) were on the top.

The seeds in (a) and (b) were in glass jars so they got plenty of light. The seeds in (c) were in a tin.

Perhaps if the seeds in (b) had been grown under the soil, they would not have sprouted first.

How could Tom make his test fair? He would have to keep *everything the same* except for the kinds of bean seeds. He could set up his test like this:

He should use the same kind of jar, the same number of beans and the same amount of soil. The only thing different would be the kind of beans in each jar. This would be a *fair test* of which bean sprouted first.

C: Example

Look again at question E on page 23 about John and his friends making tea at camp. Suppose they tried to find out which kind of cup kept their tea hottest. To make the test fair they should put the *same amount* of tea into the *same size* cups at the *same time*. Then they should leave all the cups in the *same place* before tasting the tea. Then they would be sure that the *only* thing affecting the hotness of the tea was the kind of cup it was in.

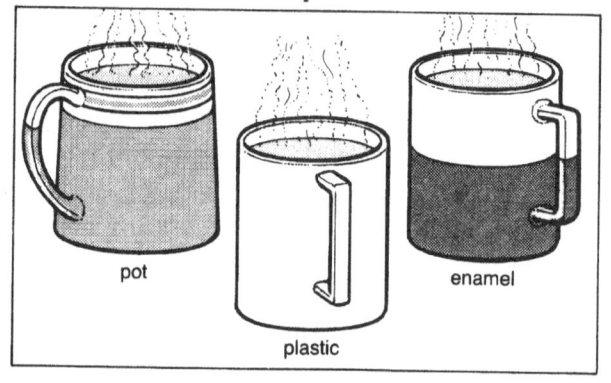

So, making a test fair means keeping everything the same except the one thing that you want to find out about.

Try Yourself

D

Look again at Sean's experiment with the nails and bottletops, on page 23. What should Sean do to make this a fair test?

E

Zoë and Keith both did an experiment to find out which kind of metal conducts heat best.

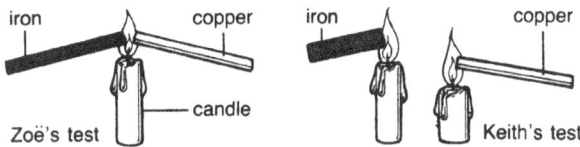

Which test is not a fair one and why?

F

Steve and Debbie were stretching rubber bands to see which were the strongest. Steve said that red ones were stronger than blue ones, but Debbie disagreed. They decided to do a test to find out which were the strongest. Which of the things in the list below would they have to keep the same to make the test fair?

1 the length of the bands
2 the thickness of the bands
3 the weights they attached
4 how long they left them for

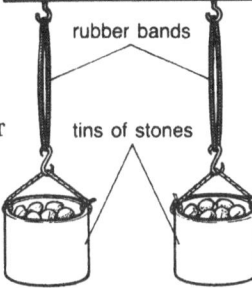

G

Look again at Sue's experiments with soil on page 23. She now wants to make a fair test to find out which soil lets water pass through the fastest. Which things will she have to keep the same in her new test?

H

Pauline wants to find out which kettle boils faster: the one on the gas cooker or the electric kettle. Which of the things in the list must she keep the same?

1 the flame on the gas cooker
2 the amount of water in each kettle
3 the make of the kettles
4 the time of starting

I

Alex did a fair test to find out which dissolves faster, salt or sugar. Fill in the missing words in his report.

I found two ..(1).. sized jars. I put the ..(2).. amount of warm water into each one. Then I added two spoons of salt to the first jar, and ..(3).. spoons of sugar to the the second jar. I stirred them ..(4).. and found that the salt dissolved first.

J

Describe how you would do a fair test to find out which person in your class has the strongest grip.

K

Owen had three different types of glue. He wanted to find out which glue was the strongest so he did a test with used matches and paperclips.

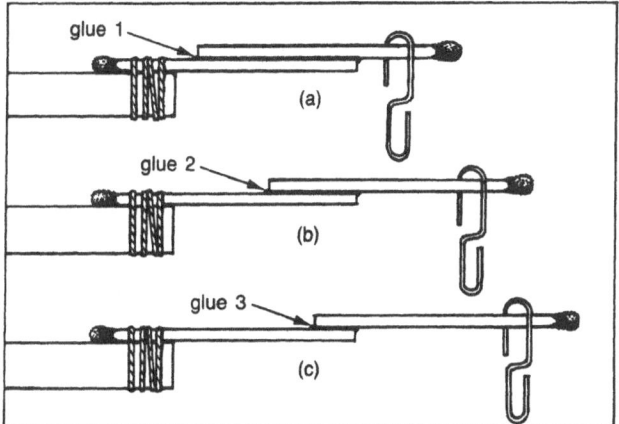

He hung small weights from the paper clips until the two matches came apart.

Study the diagrams and then fill in this table.

Things he had kept the same	Things he had not kept the same

UNIT 12
Applying ideas

In science, we find things out by doing experiments. Then we can *use* what we have found out to make new things and improve our ways of doing things. For example, after electricity was discovered, scientists made electric light bulbs, electric motors and many other things that made life easier.

You can use what you learn in science in the same way. Here are some examples.

A: Example

If you try the tests that Sue did on page 23, you will find that water drains through different soils at different speeds.

Type of soil	Drainage speed
Sandy	Fast
Loam	Medium
Clay	Slow

How can you use this knowledge?
  If you make a list of places where drainage is important, you will see that some places need good drainage but others do not.

Place	Drainage needed
Sports fields	Fast
Garden	Medium
Plant pots	Medium-slow
Fishpond	Very slow

Which kind of soil would you use to plant some seedlings in pots for your classroom?

  When you answer this question, you are *applying your knowledge* about soil drainage.
  You could test the soil on your school field and in your garden to see if it is the right kind.

B: Example

On page 13, you classified materials into those which allow heat to pass through easily (conductors) and those which do not (insulators). You can use this knowledge in several ways. For example, suppose that you put some hot tea into a plastic milk bottle. How could you keep it hot if you did not want to drink it right away?

  You could wrap something around the bottle that does not let heat pass through easily. Cloth or newspaper would be suitable.

The child's flask shown below is made of *polystyrene*. This is a kind of plastic which is a very good insulator. It keeps hot drinks hot in winter and cold drinks cold in summer.

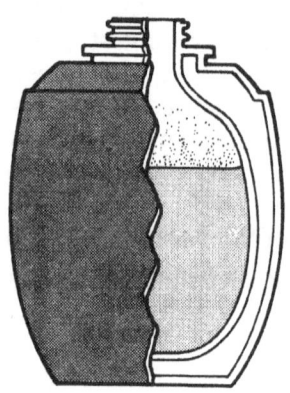

It is also important to understand *how* something works. Then you can apply this knowledge to make something else work.

C: Example

Suppose you stretch a rubber band and then let go of it. It goes back to the shape it was in the first place. This knowledge is used in many machines: catapults, model planes, holding things together and so on. We say that all these machines work on the principle that if you stretch a rubber band and then let go of it, it goes back to the shape it was to begin with.

Applying an idea **means using what you know to make something work.**

Try Yourself

D

If you stretch a rubber band over an empty tin and pluck it, it makes a musical note. The more tightly you stretch the rubber band, the higher the note will be. Make a list of all the musical instruments you can think of that use this idea.

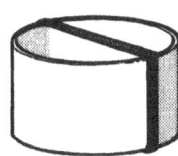

E

When a liquid is heated, it *expands* and takes up more space. When it is cooled, it goes back to the size it was in the first place. Which of these instruments uses this principle?

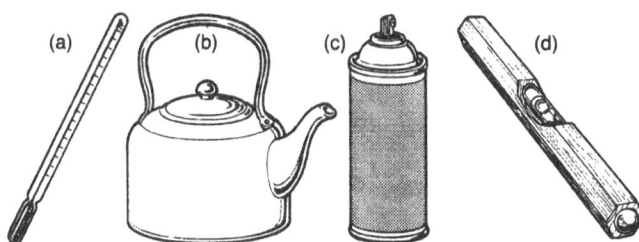

(a)   (b)   (c)   (d)

F

These three things use the same idea to make them work. What is this idea?

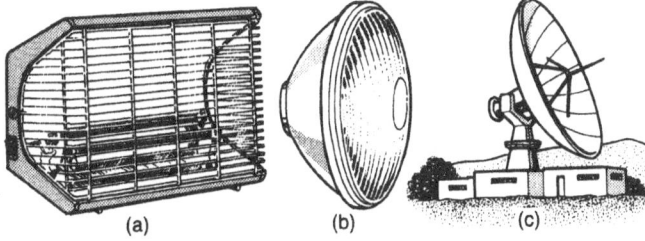

(a)   (b)   (c)

G

If you watch the smoke from a fire, you can see that hot air rises. This principle is used in *hot air balloons* and *chimneys*. Which of these machines use this principle?

1 an electric fan   3 a glider
2 a refrigerator   4 a hair dryer

H

The diagram below shows two gear wheels.

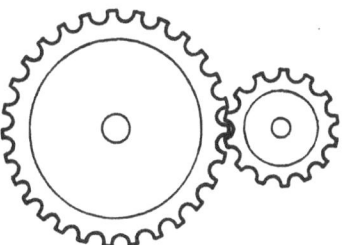

When the big wheel is turned slowly, the small wheel turns quickly. This is because the two wheels are different sizes.

Gears are used in many machines. Sometimes they are used to make things turn more quickly, and sometimes they are used to make things turn more slowly.

 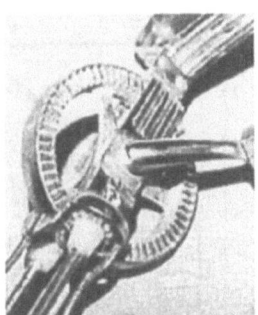

Show how these machines work by ticking the correct column in the table.

	Geared to make it go slowly	Geared to make it go quickly
Bicycle Car Clock Whisk		

I

Explain how you can use each of the following ideas:
1 When an object is placed in front of a light, it casts a shadow.
2 When water is boiled, it turns to steam.
3 Black objects get hotter than white ones when placed in the sun.
4 Most bacteria cannot live without air and water.
5 Electricity can pass through metal but not through plastic.

UNIT 13

Improvising

On page 19, you saw how Martin made a water clock from an empty plastic bottle. On page 15, there is a picture of a torch made from a battery, a bulb, a can, a strip of foil, sellotape and a rubber band. You could find most of these things in your classroom or at home.

Many of the things you need in science can be made simply from ordinary things. We call this *improvising*. You can even make a simple microscope, like this:

A: Activity

1 Cut the bottom from a matchbox tray. Then cut a 'window' in the other piece of the matchbox, like this:

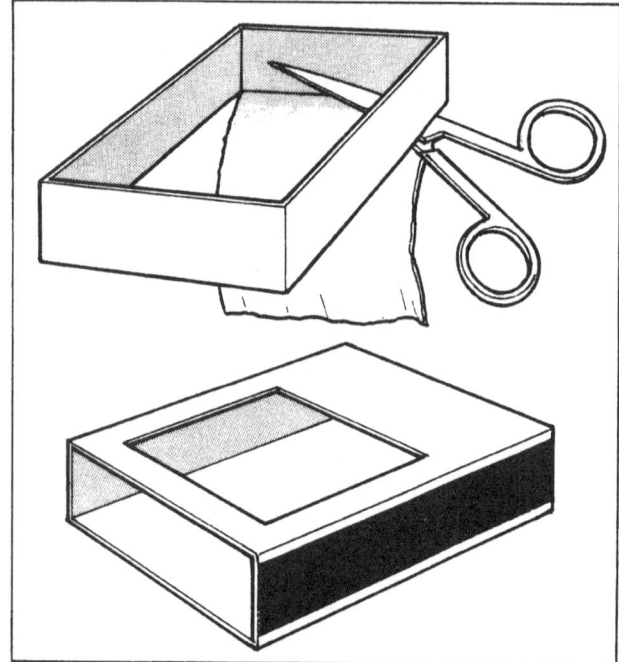

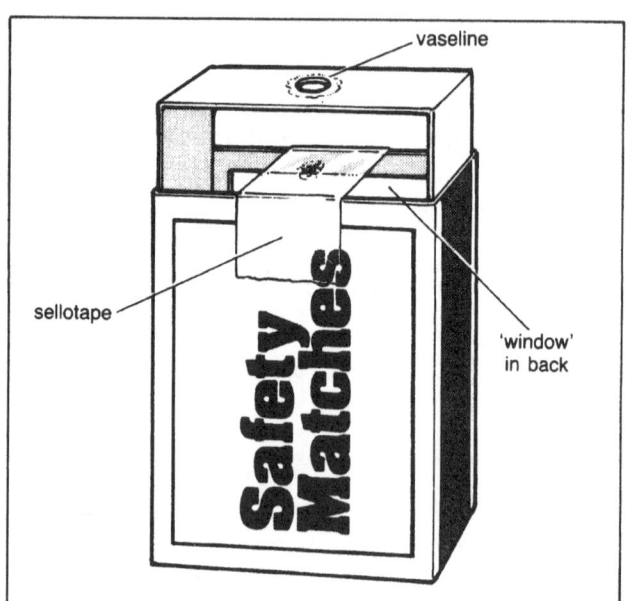

2 Make a hole in the end of the tray with a sharpened matchstick. Rub some Vaseline around the hole.

3 Put the tray back into the box with the hole and the window at the same end. Stick a piece of sellotape through the window and across the box as shown in the diagram. You should still be able to move the tray up and down.

4 Put a drop of water over the hole using a matchstick. The Vaseline will stop it from spreading. The drop of water is the *lens* which makes objects look bigger.

5 Put a tiny object on the sellotape. Hold the microscope so that the 'window' faces the light. Look down through the water drop. Move the tray up and down slowly, until the object is in focus.

There are many ways in which you can improve this simple microscope. Perhaps it would work better if you used a cigarette packet instead. Try to make a better one for yourself.

Improvising tests **is an important skill too. Do not just ask your teacher how to test something; try to find a way for yourself.**

Try Yourself

B

Look around your classroom. Try to find as many things as you can which could be used as the following:

    hammer
    screwdriver
    scoop
    ruler
    mirror
    musical instrument

C

Look at this drawing of an empty washing-up liquid bottle. If you cut round it along the dotted lines, which of the things listed below could you make?

1   a plastic cup
2   a lid
3   a funnel
4   a sieve

D

Describe how you could make a measuring jar for measuring different amounts of liquid, using these things:

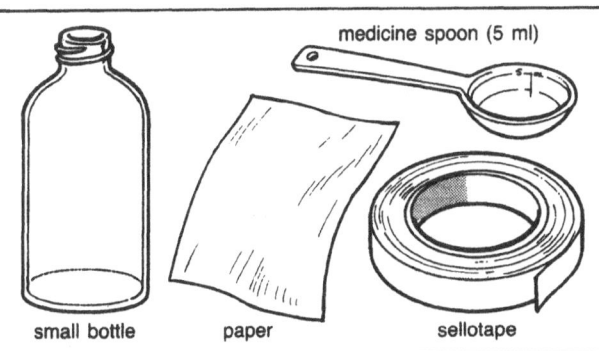

small bottle   paper   sellotape

E

Improvise a way of trapping live houseflies in your classroom. Make a list of all the things you would need. Draw a diagram to show how your trap would work.

F

This is Melanie's design for a steam turbine.

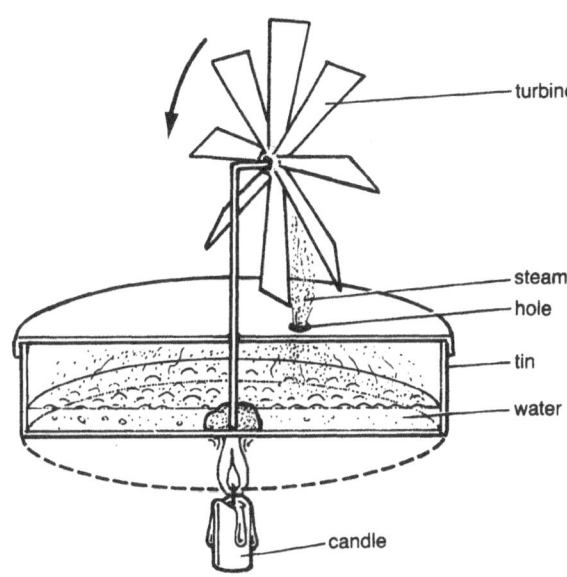

turbine

steam

hole

tin

water

candle

If you were going to build this from Melanie's design, which of these would you use to make the turbine?

1   cardboard
2   aluminium foil
3   tin sheet
4   plastic

How much water would you put in the tin?

1   enough to fill it
2   half the volume needed to fill it
3   very little

How big would you make the hole?

1   as small as possible
2   the thickness of a needle
3   the thickness of a matchstick
4   the thickness of a pencil

What else would you do to make sure that the turbine worked well?

G

Describe a test that you could improvise to answer each of these questions:

1   What is the steepest slope a worm can crawl up?
2   Do caterpillars prefer sunshine or do they prefer shade?
3   Which of these liquids evaporates fastest?
    nail varnish remover   cooking oil   water

UNIT 14

Being curious

You do not have to be curious to learn science, but it helps if you are. Curiosity makes you ask questions and observe more carefully. And when you are very curious about something, you will probably remember what you have learned much better.

Here are a few games and activities to make you curious. You can do them in groups of two or more.

A: Activity

Feely bags. Take a brown paper bag, put an object inside, and close it. Ask your friend to put a hand in the bag and find out what the object is, just by touching it. You are not allowed to look, smell or shake the bag. Take turns to be the one who 'feels' the bag.

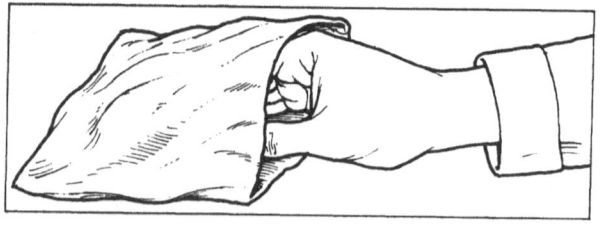

B: Activity

Noisy tins. This is a similar game, which uses hearing instead of touching. Put an object in a tin, put the lid on, and ask your partner to find out what is inside by shaking or turning the tin. You can make it harder by padding the tin or using any other tricks you can think of!

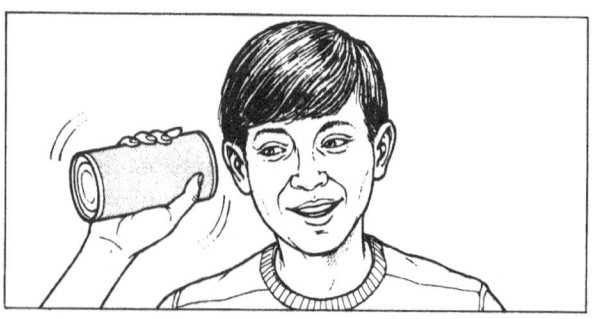

C: Activity

Blind walks. One day when you are outdoors, try a blind walk. You will need a guide, of course, so do it in pairs, taking turns to be blindfolded. The one who can see guides the 'blind' one towards interesting objects, sounds or smells. The blind person describes the things he or she finds.

D: Activity

Scavenger hunts. Your teacher might prepare a list of objects to hunt for. If not, try to find ten things which are alike in some way, such as:

ten black things
ten metal things
ten things that float
ten living things
ten things that change in water etc

E

The ten objects below are all . . . what?

matchbox
book
pedal bin
tap
daisy
pin
envelope
purse
lighter
window

Try Yourself

F

Take the strongest lens you can find and look at some familiar objects (such as a pencil point, a fingernail, your skin, a zip). Make lists of the things which you have never noticed before and the questions that come in to your mind.

G

Switch on a torch and lie it down on the floor so that its beam of light just shines across the surface. Then write down all the things you had not seen before you switched on the torch.

H

Lie down on your back on the classroom floor, and look up at the room for a few minutes. When you get up, make lists in the same way as you did in F. Try this outdoors if you can, in a field or wood or anywhere interesting.

I

In your group, have a quiz, using the questions you have listed in F and H. Take turns to ask the rest of the group a question.

J

The pictures below show familiar objects in unusual ways. What are they?

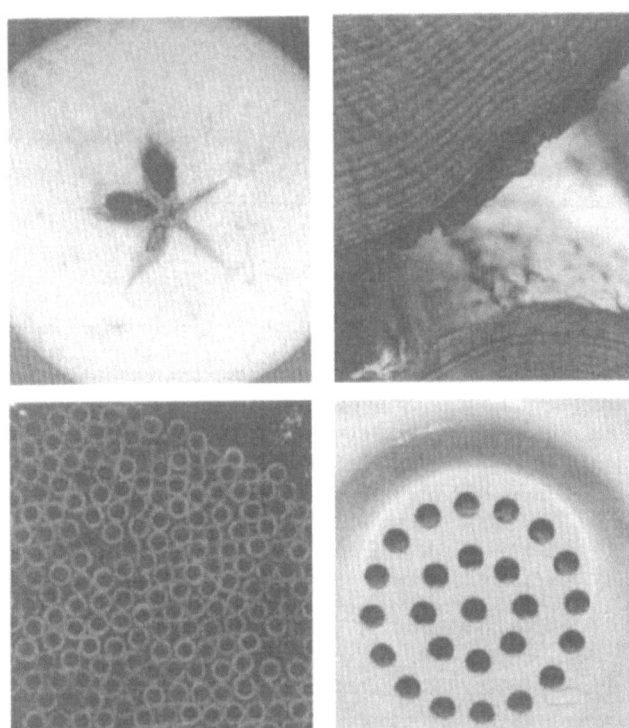

K

Take a piece of string at least a metre long, and put it down in an interesting place outdoors. Then imagine you are an ant, making a journey along the string. Get as close as you can to the string (use a lens if you want). Describe your journey, and the world you are passing through.

UNIT 15

Being practical

Being practical means doing things instead of simply sitting and listening. In science, you learn and remember things better if you have found them out for yourself by observing, experimenting, or making a test.

A: Example

In Unit 19 on p.43, there is a question about clouds, and the kind of weather they bring. You could answer this in several ways:
1   You could ask somebody the answer.
2   You could find a book about clouds, and look the answer up.
3   You could observe clouds for several days and keep records of the weather which follows them.

The most practical way is (3), because you find the answer from your own observations. The next best would be (2), because you also *do* something to find out. (1) may be the easiest way, but you may soon forget what you have learned. You can see that most of the questions in this book are about finding things out by *doing* different things.

B: Example

A group of children collected lots of plants to study.

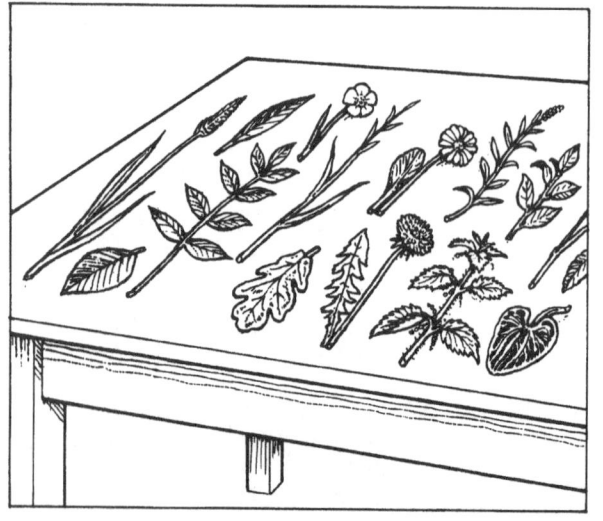

Calum wanted to find out which made the best green dye;
Moira wanted to know which plants were poisonous;
Lorraine wanted to know their names.

How did they answer their questions?

Calum crushed and heated each plant with water. Then he tested the liquids on pieces of cloth.

Moira went to a herbalist's shop where the shopkeeper told her all about the plants she had brought.

Lorraine found a book about plants in the library, and looked at the pictures until she found the names of her plants.

All three children solved their problems in different ways, but each one was being practical.

Try Yourself

C

Look at the question on matches in Unit 21 (question F, p.47). Which of these is the most practical way to answer the question?
1 reading the label on the matchbox
2 striking matches on lots of different surfaces
3 finding a book about matches
4 asking your teacher the answer

D

Describe a practical way of answering each of the following questions:
1 How does toothpaste clean your teeth?
2 What is best for removing grease marks on clothing?
3 Will sellotape stick under water?
4 Which washing-up liquid keeps its lather longest?

E

There are many ways to start a fire *without* matches or a lighter, such as using a lens, or two pieces of wood.

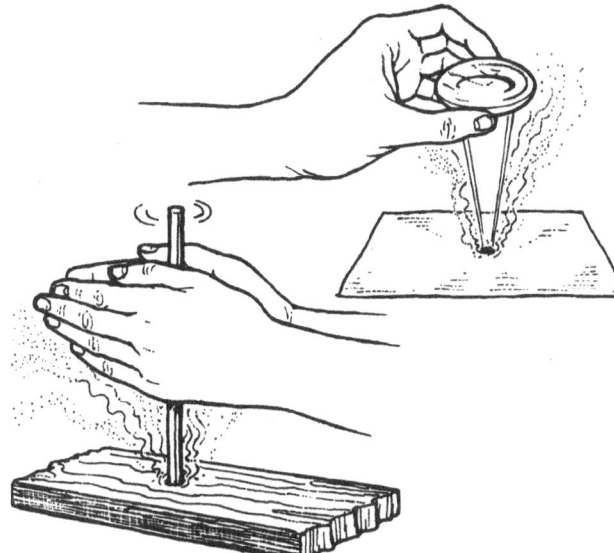

However, most of these are quite difficult unless you have learned the knack. Try *one* method out, and write down some rules for making it work successfully.

F

Make an ink mark on a strip of blotting paper, and hang it over a pencil so that the paper is touching some water in a glass, like this:

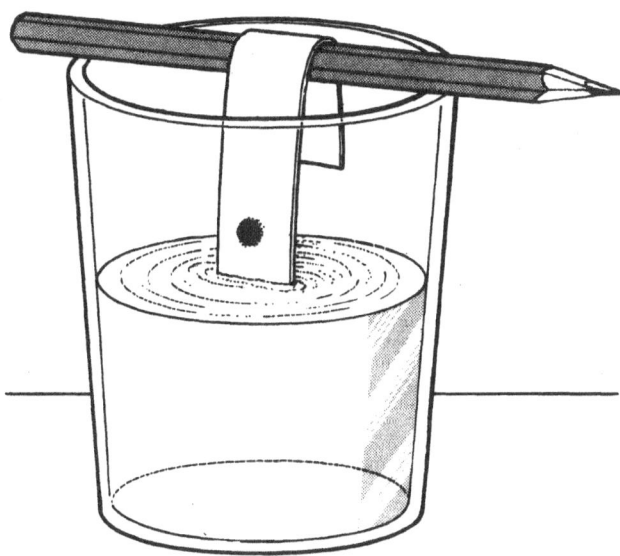

The ink will spread out and split up into different colours. Work out in a practical way how to get the best colour pattern from an ink mark.

G

What happens in F if you use other liquids in the glass instead of water?

H

What happens if you mix the ink in F with lemon juice or surgical spirit?

I

What kinds of colour patterns do you get with these 'new' inks?

J

Flour paste and egg white are both often used as home-made glues. Do some tests to find out which is the best for sticking plastic to plastic, glass to glass and metal to metal. Report your results to the class.

PART TWO: TOPICS

UNIT 16

Investigating earth

The earth is made of many materials. These include soil, rock, mud, clay, magma and sand. Each of these materials is useful to us in some way.

A

Look at these photographs, then answer the questions which follow.

(a) What is happening in each picture?
(b) In what way is the earth being useful?

B

Plants grow in *soil*. But there are many different kinds of soil, and some are better than others at helping plants to grow.

Find a place where there is some soil with plants growing in it. Look at it, touch it and smell it. Then fill in this table by putting ticks in the boxes.

My soil is:

Wet			Dry
Hard			Soft
Tiny particles			Large particles
Strong smell			No smell
Has plant remains			No plant remains

Which kind of soil do you think would be best for planting seeds in, and why?

C

Mud and clay have always been used for making pots, bricks, and houses. Look at these drawings of brick-making, and put them in the right order.

This photograph shows a volcano erupting in 1983. In your group, discuss what is happening and why it is dangerous. Then draw a diagram to explain how a volcano erupts.

D

Trevor was testing different soils to see which contained the most air. He took three glass jars, and put one cupful of soil in each:

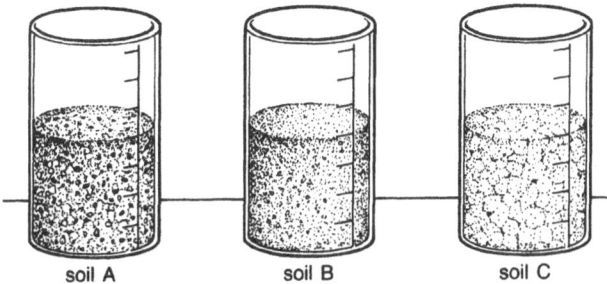

He then poured one cupful of water into each jar. These were his results:

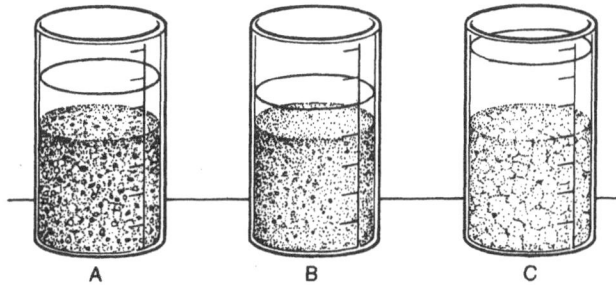

Which soil had the most air in it?

F

Many of the earth's rocks have been formed by the action of intense heat deep underground. The granite in this picture is one of these rocks.

Take some soil and bake or heat it. Find the best way of recording the difference in the soil before and after you heat it.

UNIT 17

Investigating air

Air is very strange stuff — you cannot see it, smell it, or touch it. So how do you know it is there?

A

Here is a list of some ways in which you can tell that air is all around us. How many more ways can you add to this list? Some hints are given below.
(a) blowing bubbles through a straw
(b) blowing the smoke from a candle
(c) feeling the draught under a door
(d) inflating a balloon
(e)
(f)
(g)
(h)

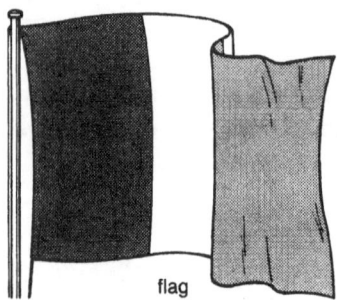
flag

dandelion clock

B

Blow up a balloon and let it go.
(a) What happens?
(b) Why does all the air come out?
(c) Why does the balloon not fall straight to the ground?

C

Light a match, then blow it out and watch the smoke. Which way does it go? Repeat the experiment several times and try to explain what you see.

D

Put your hand on your chest, then breathe deeply. Which of these describes what you feel?
(a) As I breathe in, my chest expands (gets bigger).
(b) As I breathe in, my chest contracts (gets smaller).
(c) As I breathe out, my chest expands.
(d) As I breathe out, my chest contracts.

E

Breathe out gently on to a cold surface, such as a window or a metal object. What do you see?
   This shows that the air you breathe out contains one of these substances, which one?
(a) oxygen
(b) carbon dioxide
(c) water vapour

F

Here is a glass of tap water that has been standing on a windowsill for a few hours.

(a) what can you see in the water?
(b) where have they come from?
(c) explain the importance of what you have observed.

G

The *atmosphere* in which we live and breathe is made of air several miles deep. This *presses down* on us, although we cannot usually feel it. However, we *can* create some situations in which we *can* feel this pressure.

We can see and feel air pressing on water in a glass, like this:

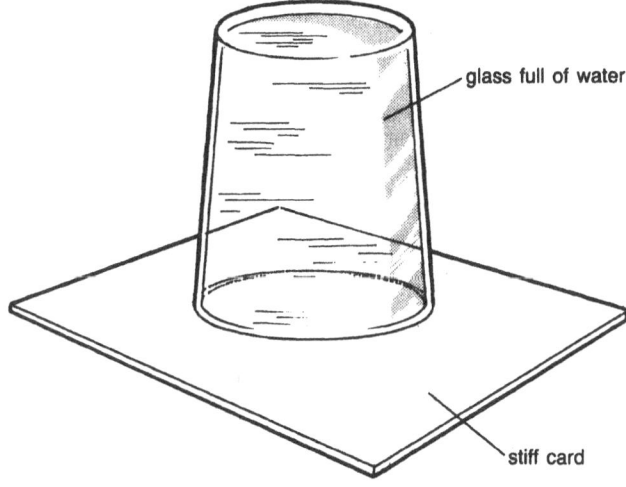

glass full of water

stiff card

You can create suction, like this:

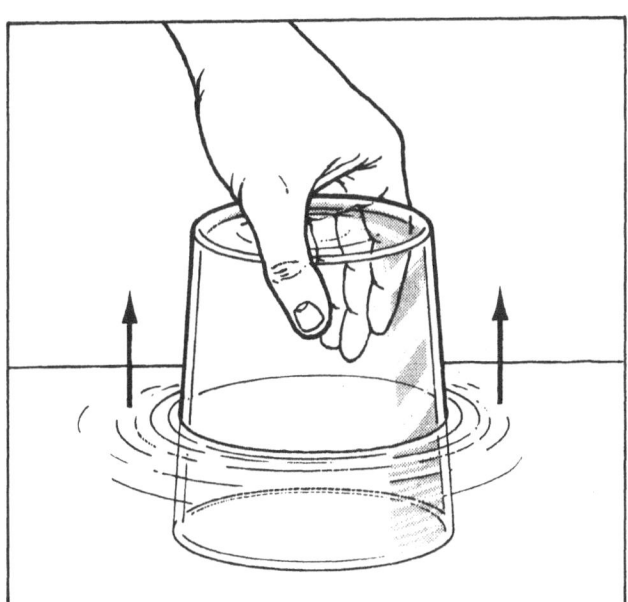

There is *most* suction when the glass is:
(a) full of water
(b) empty
(c) half full of water
(d) almost out of the water
(e) right underneath the water

H

How great can you make the air pressure in your lungs? Find a *practical* way of comparing your lung pressure, or 'blowing power', with the others in your group. (*Hint*: you might use a balloon or a long piece of plastic tubing.)

I

Air pressure is measured by a *barometer*. Is there one in your school or your home? Ask your teacher to explain it to you.

UNIT 18

Investigating water and other liquids

A liquid is a substance which you can pour, and which takes the shape of whatever container you put it in.

A

Put a circle around each of the liquids in this list:
Plastic Oil Milk Gravel Paper Vinegar
Lava Sand Tea Jam
Discuss those that you are not sure about.

B

Liquids often look alike, so sometimes you can mistake one for another. This can be dangerous because some liquids are poisonous, and some can hurt your skin.

Here are some ways to distinguish between liquids:
(a) What colour are they?
(b) How easily do they *pour*?
(c) How do they *feel* (greasy, sticky, cold, etc)?
(d) How do they *smell*?
(e) How easily do they *dissolve* substances?
Compare some liquids which you have at school or at home (such as water, cooking oil, white spirit, vinegar), and then make a table like this:

	Water	Cooking oil	Spirit
Colour			
Feel			
Smell			
Ease of pouring			
Dissolves salt			

Always be careful when *touching* or *smelling* liquids.

C

Most of the water in rivers, lakes and seas is unsafe to drink, because it contains *bacteria* from people and animals. Even water without bacteria can look and taste bad because of other substances dissolved or mixed in it.

In this country, water is *purified* by removing the germs and other substances. Germs are killed by adding *chlorine* (you can smell it in swimming-pool water). Other solids are removed by *filtering*. A filter is shown in the picture below.

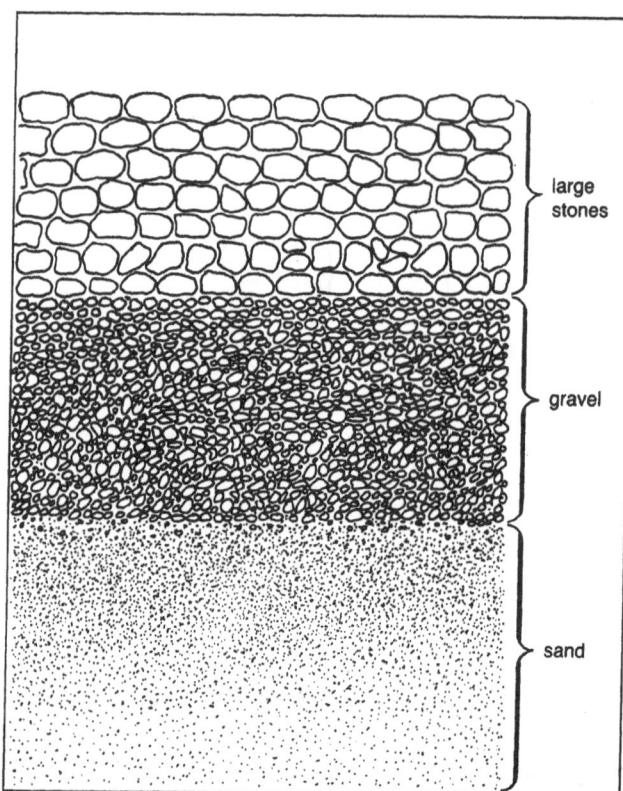

large stones

gravel

sand

Why are there large stones at the top and sand at the bottom?

D

Take some dirty water from a puddle, ditch or pond, and pour it through some cotton wool. Look at the cotton wool and the water afterwards. What changes can you see?

When Marie did the same thing, this is what she saw:

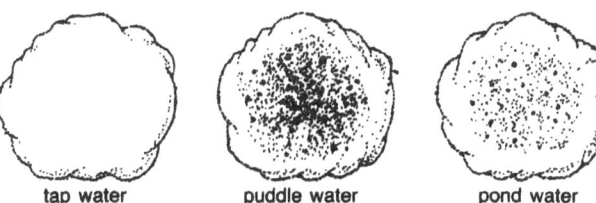

tap water   puddle water   pond water

What was her conclusion?

E

Suppose you did not have a supply of clean water at your school. Draw a diagram to show how you could make a simple water-filter using an *oildrum*, *stones*, *pebbles* and *sand*. Make a list of any other things you would need to make the filter work well.

F

Graham put some ice cubes in a glass and measured the temperature every five minutes. These were his results:

TIME		TEMPERATURE
0 MINUTES		0°C
5 "		0°C
10 "		0°C
15 "		1°C
20 "		4°C
25 "		8°C

Explain in words what Graham's chart tells you.

G

When a liquid is warmed, it *evaporates* (turns into a gas). The hotter it is, the faster it evaporates. When it is very hot, it *boils*.

All of the things below are caused by evaporation. Can you add some more to the list?
(a) Lakes dry up in summer.
(b) A tin of paint goes hard if the lid is left off.
(c) People become thirsty after hard exercise.
(d) Porridge burns if it is cooked for too long.
(e) Pans are hard to clean when left for a long time.

H

Ask at home if you can put some cooking oil and some other liquids into the freezer. Leave them for a day and then observe what has happened. Did they freeze like water? How did they change?

UNIT 19

Investigating weather

Wherever you are, you can observe the weather and how it changes. You can even learn to *forecast* the weather, at least for the next few hours.

Weather changes because air moves around the earth. In Britain, it is usually true that:
(a) air from the north is cold
(b) air from the south is warm
(c) air from over the sea is wet
(d) air from over the land is dry

A

Look at the map of Britain and its surroundings below.

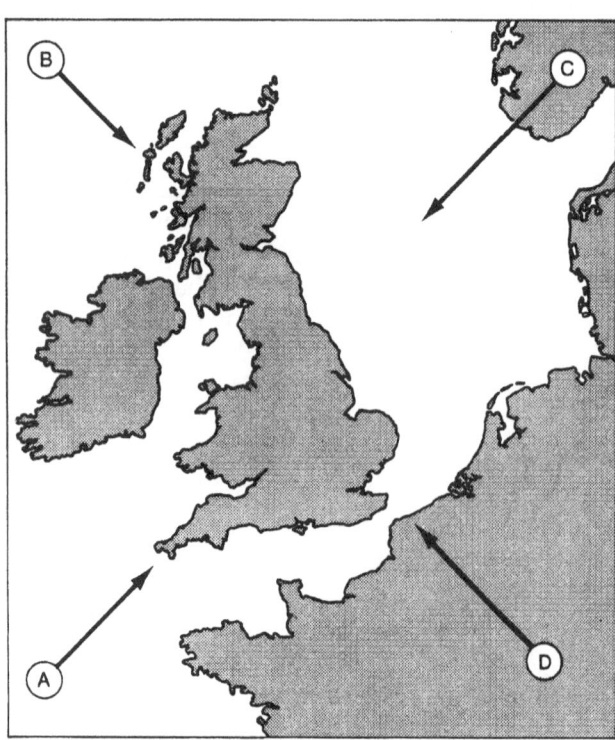

You can see that if the air is coming to Britain from the south-west (A on the map), it is coming over the *sea*, so it will be *mild* and *wet*. This is the most common wind direction in Britain, which is why our weather is so bad!

Now write down the kind of weather you would expect if:
(a) the wind was from the north-west (B) in winter
(b) the wind was from the north-east (C) in spring
(c) the wind was from the south-east (D) in summer

B

You are in the playground on a cloudy, dull day, and you notice from the weather cock that the wind is changing from south-east towards south-west. How do you think the weather is going to change?

C

Some children made rain gauges and put them in the school playground in the places shown in the picture. Which have been put in bad places? Why are these places unsuitable?

D

Here are some pictures of different types of clouds.

What kind of weather do you usually get when you see these clouds?

E

Some people say, 'Red sky at morning, shepherds' warning'. What do you think it means?

In your group, make a list of all the sayings you can think of which are about the weather.

F

The two charts below show how much rain fell each month last year in London and Nairobi.

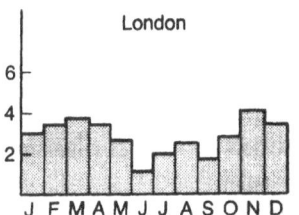
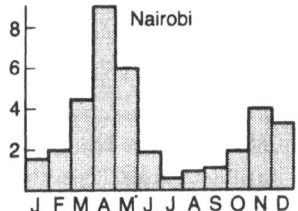

(a) Which was the *wettest* month in *London*?
(b) Which was the *driest* month in *Nairobi*?
(c) Which city had the *most rainfall* in the year?
(d) If you were a farmer living near Nairobi, when would you plant your crops?

G

Look at this chart from a daily newspaper, then answer the questions.

AROUND THE WORLD
Lunch-time reports

		C	F			C	F
Ajaccio	S	24	75	Luxembourg	F	15	59
Algiers	S	24	75	Madrid	S	31	88
Amsterdam	F	14	57	Majorca	S	25	77
Athens	S	24	75	Malaga	S	26	79
Bahrain	S	34	93	Malta	S	27	81
*Barbados	C	29	84	Manchester	F	14	57
Barcelona	S	24	75	*Mexico C	F	23	73
Belgrade	C	19	66	*Miami	F	30	86
Berlin	F	16	61	Milan	C	25	77
*Bermuta	F	27	81	*Montreal	S	30	86
Biarritz	F	21	70	Moscow	F	21	70
Birmingham	F	13	55	Munich	Th	15	59
Bordeaux	S	21	70	Nairobi	C	22	72
*Boston	S	22	72	Naples	S	25	77
Boulogne	F	13	55	Newcastle	F	15	59
Bristol	F	16	61	*New York	S	31	88
Brussels	F	14	57	Nice	S	25	77
Budapest	C	23	73	Oporto	S	30	86
Cairo	S	35	95	Oslo	F	14	57
Cape Town	S	18	64	Paris	F	17	63
Cardiff	F	16	61	Peking	S	33	91
Casablanca	S	27	81	Perth	C	21	70
*Chicago	F	30	86	Reykjavik	C	6	43
Cologne	C	16	61	Rhodes	F	24	75
Copenhagen	F	17	63	*Rio de Jan	S	26	79
Corfu	Th	21	70	Riyadh	S	41	106
Dublin	F	15	59	Rome	S	26	79
Dubrovnik	F	22	72	*San Fran	S	16	61
Edinburgh	S	14	57	Salzburg	C	12	54
Faro	S	31	88	*Sao Paulo	C	23	73
Florence	C	26	79	Seoul	F	29	84
Frankfurt	F	13	55	Stockholm	F	20	68
Funchal	S	26	79	Strasbourg	F	16	61
Geneva	S	19	66	Sydney	C	15	59
Gibraltar	S	24	75	Tangier	S	30	86
Glasgow	F	14	57	Tel-Aviv	S	32	90
Helsinki	C	22	72	Tenerife	S	32	90
Hong Kong	F	30	86	Tokyo	S	27	81
Imsbruck	C	12	54	*Toronto	S	30	86
Inverness	C	12	54	Tunis	F	27	81
Istanbul	S	25	77	Valencia	S	25	77
Jo'burg	F	16	61	*Vancouver	F	18	64
Las Palmas	S	28	82	Venice	F	24	75
Lisbon	S	33	91	Vienna	F	18	64
Locarno	F	25	77	Warsaw	C	17	63
London	F	17	63	*Washington	F	33	91
*L Angeles	S	20	68	Zurich	F	17	63

C, cloudy; F, fair; Fg, fog; R, rain;
S, sunny; Th, thunder.
*Previous day's reading.

(a) Which was the hottest place?
(b) Which places had thunder?
(c) Which place had the most sun, London, Manchester or Edinburgh?
(d) What time of year do you think this chart was made?

UNIT 20

Investigating natural habitats

Habitats are the surroundings in which plants and animals live. For example:
(a)   Worms live in the soil.
(b)   Pandas live in bamboo forest.
(c)   Penguins live on the rocky islands of the Antarctic Ocean.

A

There are many different habitats around your home or school. Look at the photographs below. Do you have any similar habitats near your home or school? Explore one of them and find out what plants and animals live there.

B

Find a place which is always in the sun, and another place which is always in the shade. See if you can find a plant or animal which prefers living in one to living in the other.

C

Many living things have had their habitats destroyed by people, as this table shows.

Things people have done	Living things affected
Uprooting hedges Polluting rivers	Mice, birds, flowers Fish, water plants

Find out some more things to add to the table.

D

Joanna found lots of small red eggs growing on a leaf in her garden, and did this drawing of them.

Write down three questions that Joanna could have asked about these eggs. Then describe some *practical* ways of finding answers to her questions.

E

In most habitats, many of the living things are dependent on each other. They may also be dependent on some of the non-living things in the habitat. For example, in a stream you may find otters which eat fish; the fish eat snails and flies; the snails need chalk and eat algae; the flies live amongst the green plants and stones for shelter.

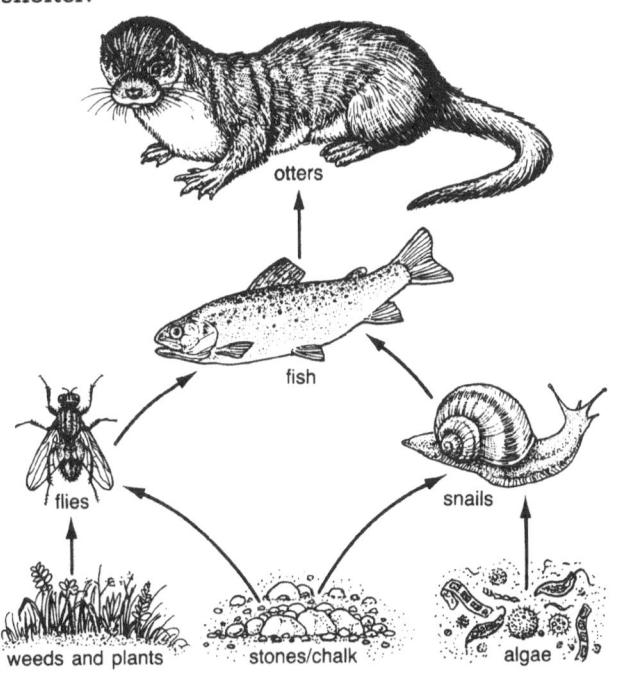

How would the changes listed below affect the habitat?
(a) All the weeds are removed.
(b) Chalk is removed from the stream bed.
(c) Chalk is added to the water.
(d) The otters are killed.

F

In your group, organise a worm hunt. Decide *where* to look, *when* to go, how to *record* what you observe, and what you need to *collect* some worms.

Write a plan for the hunt *before* you start. When you get back, report on the success of your hunt to the other groups, and suggest how your worm hunting could have been improved. And if you cannot find any worms, organise a hunt for something else —flies, spiders, caterpillars, etc.

G

Do you think worms can live in your classroom? The answer is 'yes', if you provide the right habitat.

Design a 'wormery' for your classroom. It should be possible for you to observe the worms living in it. Make a *list* of what you would need to make it. Draw a *diagram* to show what it would look like.

If you have the time and materials, try to make the wormery and see if it works.

H

Put these animals into groups which live in the same kinds of habitat:

rabbit	gull
monkey	mole
seal	crab
woodpecker	kingfisher
bluetit	koala bear
fox	caterpillar
penguin	badger

UNIT 21

Investigating the man-made environment

A

People have always built houses to live in: here are four kinds of houses which people still use today.

Compare the houses, then fill in this table.

Type of house	Advantages	Disadvantages

B

All kinds of animals have made their homes in people's houses: spiders, mice, sparrows and houseflies are some examples. Why should this be?

C

Which of these ways is used to heat your home?
- (a) coal fires
- (b) electric fires
- (c) oil central heating
- (d) solid fuel central heating
- (e) gas fires
- (f) night storage heaters
- (g) gas central heating
- (h) under floor heating

Ask all the others in your class, and then make a chart to show which is the most common.

D

This diagram shows how the Romans used to heat their villas. Write a description of how it works.

fire

floor supports

E

Pens have changed and improved in many ways since the quill pen being used here.

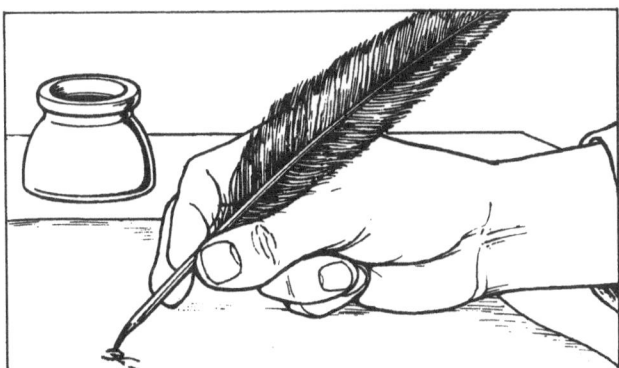

During the next few weeks, make a class display of 'Pens through the ages'. Show what they were made of and how they worked.

F

Matches are an easy way to start a fire, but can be dangerous if not handled carefully.

Make a table like this to show which surfaces a match will strike on. Each time you strike a match, blow the flame out at once. Then dispose of the match *carefully*.

Things matches will strike on	Things matches will not strike on

What conclusion can you draw from these results?

G

Look around your area to find something that is being constructed — a house, a road, a pipeline, scaffolding, etc.

Observe what is happening. Ask the people working on the construction what they are doing and how they are doing it. *Be careful as you approach the construction site. Ask permission before you enter it.*

H

Search the area around your home or school for as many different types of *walls* and *fences* as you can find. Then answer these questions:
- (a) What materials are used to make them?
- (b) How high are they?
- (c) Which kind are the strongest, and which the weakest?
- (d) What are they for?
- (e) Who is responsible for them?

I

Imagine you are planting a small garden, and you want to put a fence round it to keep out cats and dogs. The fence needs to be 50 metres long.

Decide which kind of fence you would use, then work out how much material you would need to build it.

UNIT 22

Investigating reproduction

All living things reproduce. In most cases a female's egg cell has to meet a male's sperm cell before a young plant or animal will develop. This meeting of egg cell and sperm cell is called *fertilisation*.

In the first picture, a male salmon is fertilising the female's eggs by squirting a milky substance over them.

The second picture shows a human *sperm* (from the male) entering the female egg. A baby will begin to grow from these two joined cells.

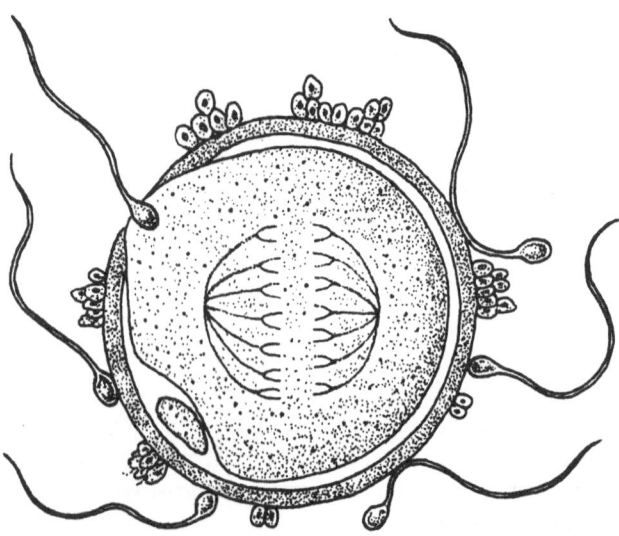

A

Why do you think that the female salmon lays hundreds of eggs, when a woman usually only produces one egg at a time?

B

Many plants are fertilised with the help of bees. Find out how this works.

C

This picture shows some hens' eggs in an *incubator* (a warm box where they can hatch). You can see that some have already hatched and some have not.

Explain how you could find out the average length of time it takes for an egg to hatch.

D

Whenever you can, observe a newly born kitten, puppy, rabbit, baby or other mammal, and make a table like this.

Things it can do when born
Things it cannot do when born
How long it takes to learn to do these things

E

The pictures below show a human *embryo* (a baby before it is born) at different stages.

When you have studied the pictures carefully, write a short description of how the embryo develops. Say which parts appear first, how its shape changes, and so on.

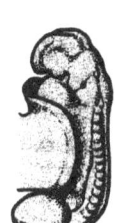

3 weeks
7 somite

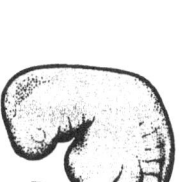

3½ weeks
14 somite

4½ weeks
5 mm

Late 6th week
13½ mm

6½ weeks
17 mm

8½ weeks
30 mm

F

The picture below shows a human baby being born. Find out what these words mean:
(a) labour
(b) contractions
(c) natural childbirth
(d) umbilical cord
(e) afterbirth
(f) midwife
(g) breach delivery

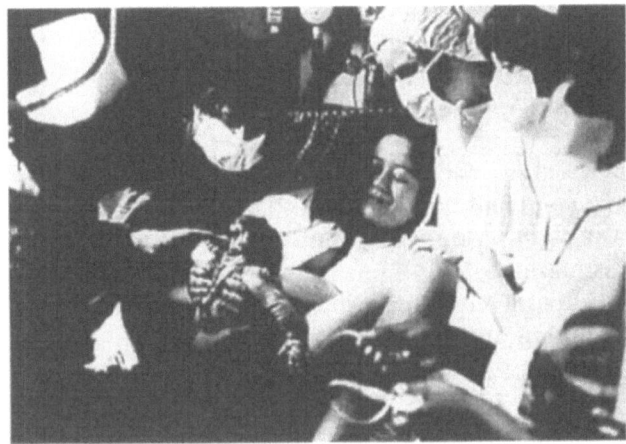

UNIT 23

Investigating food

Food is important because
(a)   it is the *fuel* which gives you energy;
(b)   it *builds* and *maintains* your body parts;
(c)   it *regulates* (controls) the way your body works.
Different kinds of foods (*nutrients*) help you in different ways:

To stay healthy, you need to eat *enough* food of the *right kind*. If you don't, you may suffer from *malnutrition*.
   Having a *balanced diet* means eating the right amounts of the three main types of nutrients.

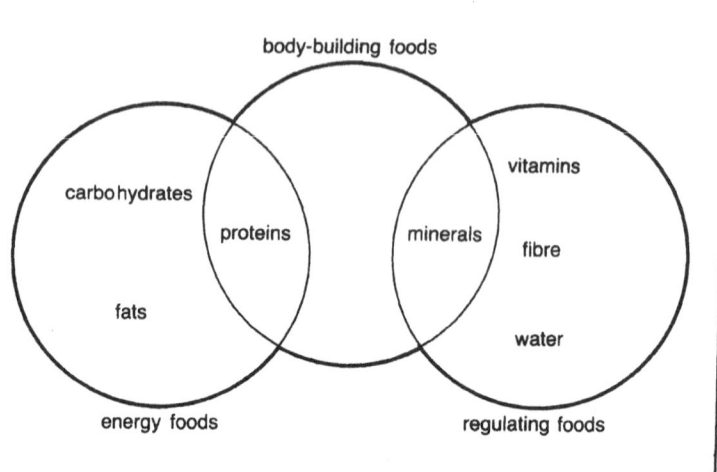

A

This table shows you the nutrients that some foods contain.
Look at the table and decide which of these sentences are true and which are false.

(a)   Bread contains hardly any fat.
(b)   Cheese contains a lot of fat and protein.
(c)   Raisins are rich in vitamins.
(d)   Apples and oranges have no protein.
(e)   Vitamins are found in most of the foods.

	Carbohydrate	Fat	Protein	Minerals	Vitamins	Fibre
Bread	√√		√		√	√
Cheese		√√	√√	√	√	
Eggs		√	√		√√	
Milk	√	√	√	√	√√	
Fish			√√		√	
Meat (Beef)		√√	√√	√	√	√√
Liver		√	√√	√	√√	
Baked Beans	√√		√	√	√	√√
Carrots				√	√	√√
Peas	√				√√	√√
Apples	√					√√
Oranges	√			√	√√	
Raisins	√√			√		√√

√√ contains a lot   √ contains a little

B

Put the foods listed below into the right sets in the diagram;

(a) raisins   (f) carrots
(b) bread   (g) liver
(c) eggs   (h) apples
(d) cheese   (i) baked beans
(e) oranges

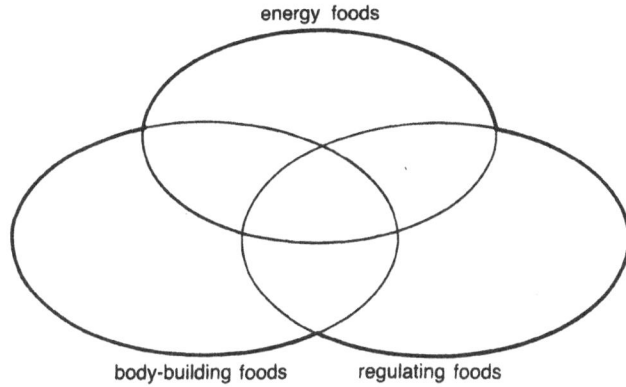

energy foods

body-building foods   regulating foods

C

The *amount of energy* which food gives is measured in *calories* or *kilojoules*. Sometimes food labels tell you how many calories the food contains. Look at this table from a cereal packet.

	INGREDIENTS 100% whole wheat	
✓	Natural wholewheat with wheatgerm and bran retained	
✓	Source of natural fibre	
✓	Energy value 1505 kilojoules per 100g (350 calories per 100 g)	
✓	NO ADDED SUGAR	
✓	NO ADDED SALT	

An average, 10 year-old child needs about 3000 calories a day to stay healthy. If you ate nothing but this cereal, how much would you need to eat?

D

Look at different cereal packets in a shop and find out

(a) which nutrients they supply
(b) which kind gives the most calories for the same weight

E

Millions of children in Africa and Asia have less than 2000 calories a day and often there is no protein in their diet.

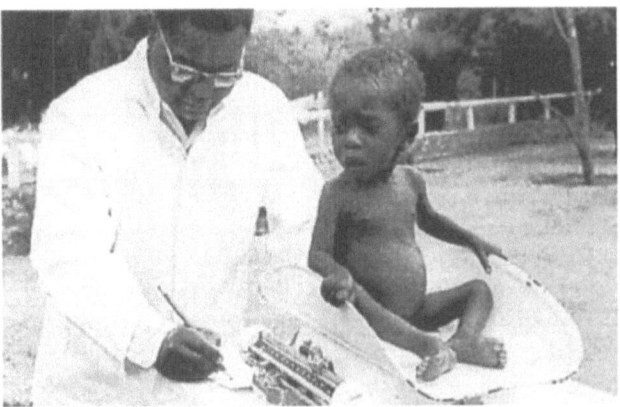

Think of some ways in which this could be prevented and discuss why it is difficult to get rid of malnutrition in the world.

F

Most food *decomposes* (goes bad) if it is not preserved in some way. Decomposition is caused by bacteria which live on the food.

On the left are some ways of preserving food. Match them up with the foods listed on the right.

By *heat* (cooking, pasteurising)   Onions
By *cold* (freezing)   Peaches
By removing *water* (drying)   Milk
By removing *air* (sealing,   Fish
canning)   Meat
By *acid* (pickling)   Grapes

G

If you cut or bite an apple and leave it, it usually goes brown. Try to stop this from happening by using one of the ways of preserving listed in F.

UNIT 24

Investigating health

To be healthy, you need all these things:
(a)   a balanced diet
(b)   clean and safe surroundings
(c)   dry and warm housing
(d)   enough sleep
(e)   enough fresh air
(f)   enough exercise
If any of these are missing, you are more likely to get sick.

A

Find out what time each person in your group goes to bed and gets up. Then work out the average amount of sleep each person has per night.

B

Ask your teacher for a map of the area round your school. Mark on it all the places where there is something *dangerous* (such as an unfenced stream) or *unclean* (such as a rubbish tip).

Use a key or code to show which are *very* unhealthy, *slightly* unhealthy and so on.

C

Some illnesses are *infectious*, which means they spread from one person to another. Other illnesses, such as asthma, cancer, anemia, fits, are not infectious.

Infectious illness	*How it spreads*
colds, coughs, 'flu	By coughing, sneezing, flies, etc.
measles, mumps, chickenpox	By coughing, sneezing, flies, etc.
diarrhoea	Dirty hands, water, flies
warts, ringworm, impetigo, lice, fleas	Touch Touch, clothes

Make a list of things you could do to avoid spreading (a) a cold, (b) diarrhoea, (c) lice.

Illness	How to prevent it spreading
(a) a cold (b) diarrhoea (c) lice	

D

This sign is often seen in food shops.

Make a list of other precautions which are taken to keep food clean in shops.

E

Millions of children in the world die every year because they use bad water and poor sanitation. This drawing shows how to make a safe lavatory.

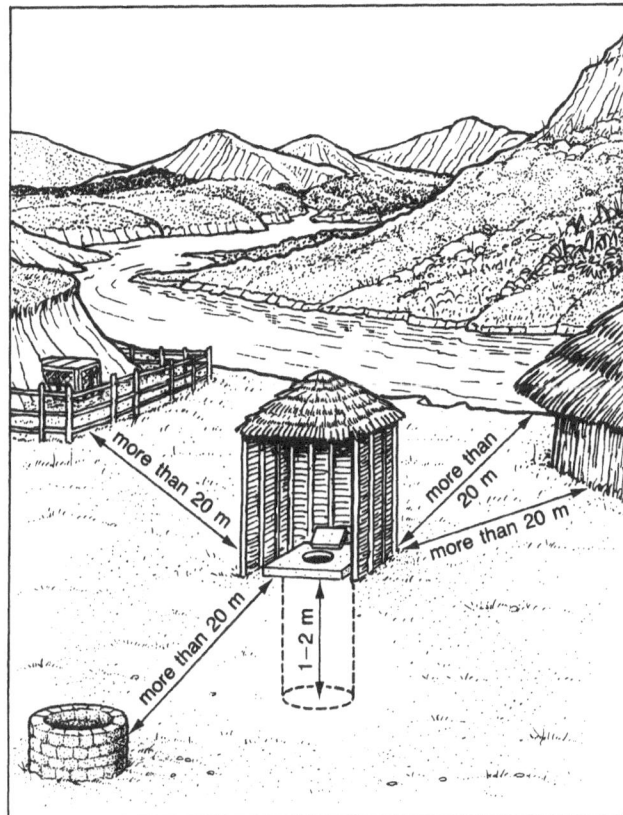

Study the picture, then explain:
(a) why it needs to be so deep
(b) why it needs a lid
(c) why it must be so far from buildings, animals, wells, and the river

F

Group the illnesses below into those which need treating by a doctor and those which can be treated safely at home by you and your family:
(a) mumps
(b) chickenpox
(c) pneumonia
(d) indigestion
(e) boils
(f) earache
(g) deep, dirty cuts
(h) stings
(i) sprains

G

Ask your teacher to show you the First Aid Box in your school, and where it is kept. Find out what each of the things inside is used for.

H

When you need medical help of some kind, it is important to know the right place to go.

Where would you go in your area for each of these?
(a) to get medicine on prescription
(b) to get a bandage for a cut
(c) to get treatment for a friend with a broken ankle
(d) to ask the district nurse to visit

UNIT 25

Investigating work

Doing work means using energy to lift, move, carry or stop something – including yourself.
When you ride your bike up a hill, you are working hard to move you and your bike forwards and upwards at the same time. When you come downhill, you are doing work when you put the brakes on.

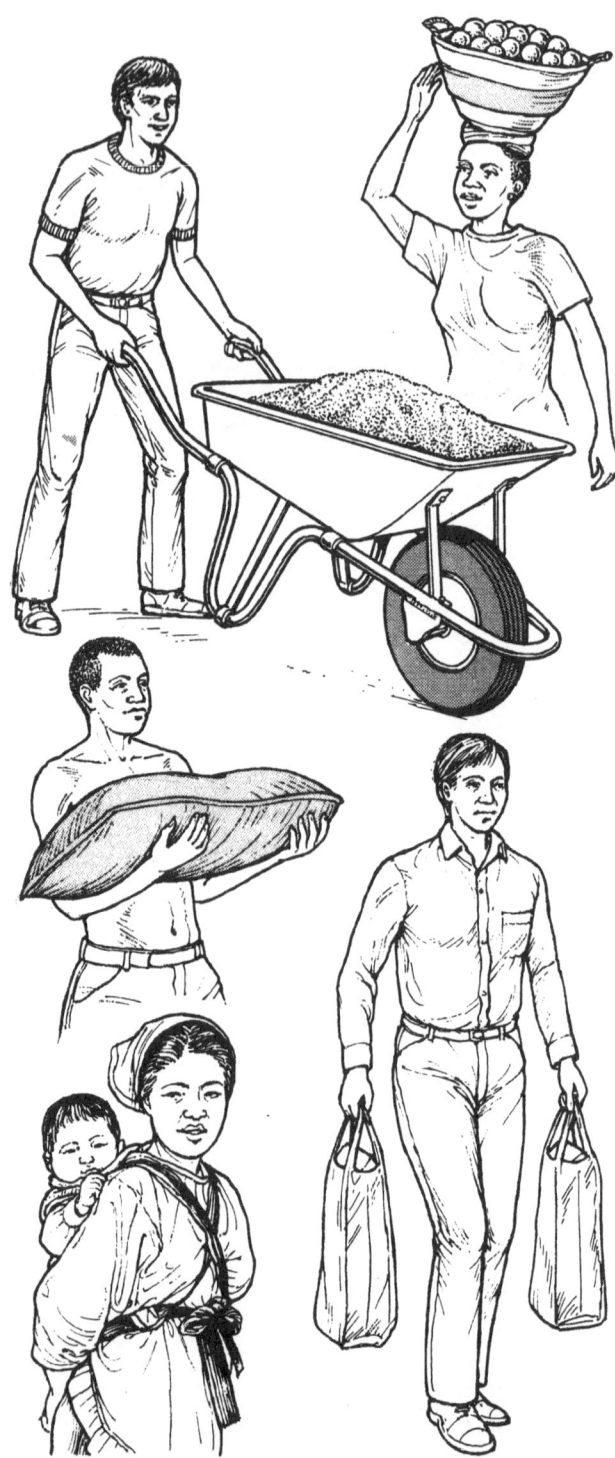

A

The more you use your muscles, the harder you are working. Which of each pair is the hardest work?
(a) climbing:running
(b) sleeping:swimming
(c) digging:writing
(d) climbing stairs:wheeling a barrow

B

Energy for driving machines can come from many sources, for example, the wind, water, steam, petrol, electricity. Write down the sources of energy that can be used to make these machines work.
(a) a bike
(b) a train
(c) a boat
(d) a satellite
(e) a flour mill
(f) a water pump

C

Some ways of lifting and carrying are easier than others. Look at the pictures opposite.

(a) Which of these ways would you choose to carry a heavy load a great distance?
(b) Which way of carrying is the hardest work?
(c) Explain why you chose these answers.

D

When Stonehenge was built, there were no cranes. Draw a picture to show how these big stones could have been stood on top of each other.

E

Tools and machines are things that people have invented to make work easier.

Some of them, like the hammer, use only muscle energy. Some, like an electric fan, use other sources of energy. Others, like the vacuum cleaner, use both muscle energy and another source of energy.

Put these things into the right sets:

Set A: muscle energy only
Set B: other sources of energy
Set C: muscle energy and another source of energy

(a) screwdriver
(b) blender
(c) power drill
(d) whisk
(e) pliers
(f) hair dryer
(g) lawn mower
(h) spanner
(i) cooker

F

Make a list of tools and machines which help to make these kinds of work easier.
(a) cutting down trees
(b) smoothing wood
(c) joining pieces of wood
(d) digging holes to plant trees
(e) picking fruit from trees

G

The chart below shows the work done in one day by a typical woman in an African country.

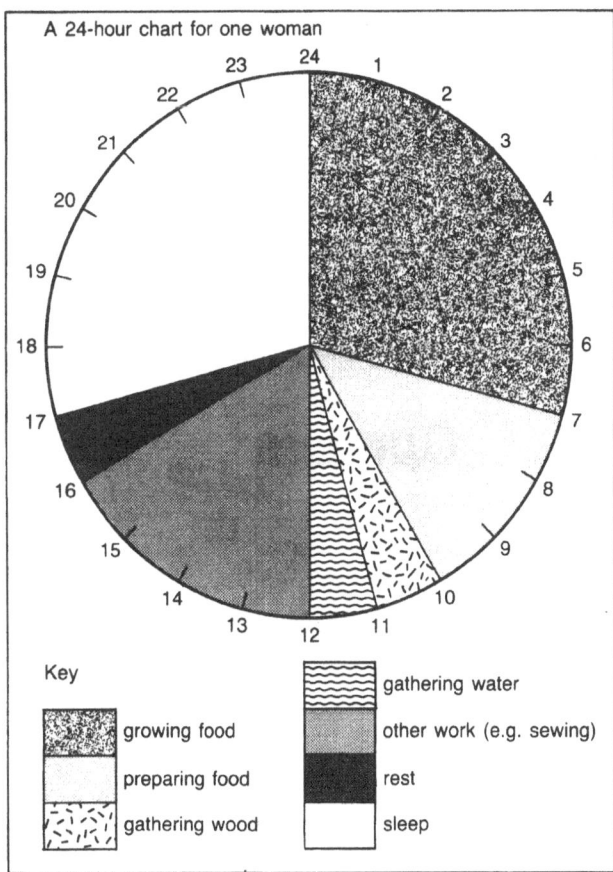

Make a similar chart for the work done by your mother in a day.

Compare the two charts, then answer these questions:
(a) What work does the African woman do which your mother does not do?
(b) What work does your mother do which the African woman does not do?
(c) How could their day's work be made easier?

UNIT 26

Investigating tools

A

Alison and Gary were sorting out a pile of tools into boxes for their grandad. Which are in the wrong box?

Where do they put the ones on the floor?

Turners

Openers

Cutters

B

Look around your house or classroom and find as many things as you can with *handles*.

Handles have different uses. Look at your list, and then discuss in your group what handles are for, and how they make work easier.

C

Tracey's mum was screwing a shelf bracket to a wall, like this:

Which of these screwdrivers would be the best to use?

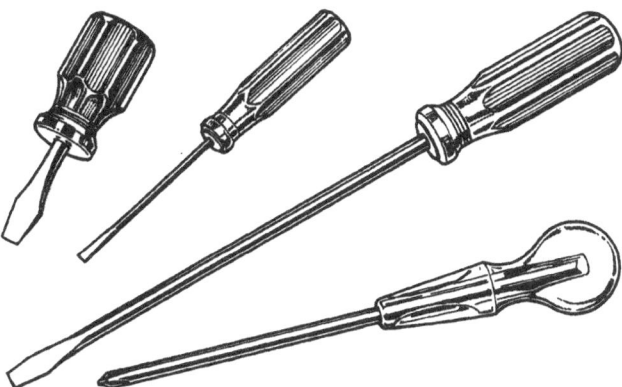

D

A *lever* is a tool which makes lifting easier. You move one end down and it lifts the other end up, like this:

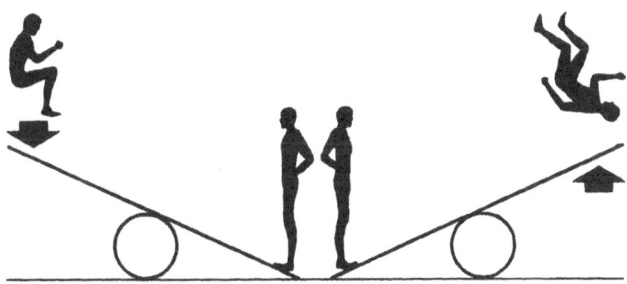

Many tools are levers in disguise. Here a spoon is being used as a lever to lift a lid.

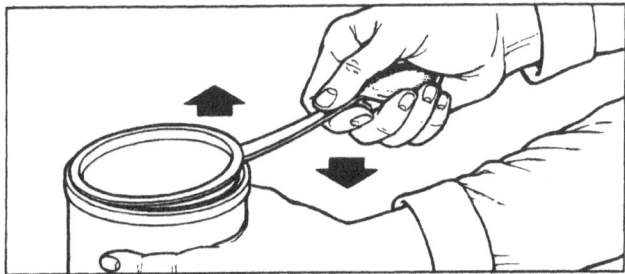

Think of more examples of common situations in which you use levers.

E

You have to move a large rock a distance of about 2 metres. The rock is much too heavy to lift. What simple tool could you make or improvise to help you move the rock?

Draw a diagram to show how you would use the tool.

F

You can do lots of experiments with a seesaw. If you do not have one nearby, make one with a plank and a big drum or stone. Then try some of these:

(a)   Who goes up?

(b)   Who goes down?

(c)   Where must you sit to balance?

G

Here are some drawings of levers. Copy them out and on each one, draw *one arrow* to show where you would push to make it work. Then write down what each one can lift.

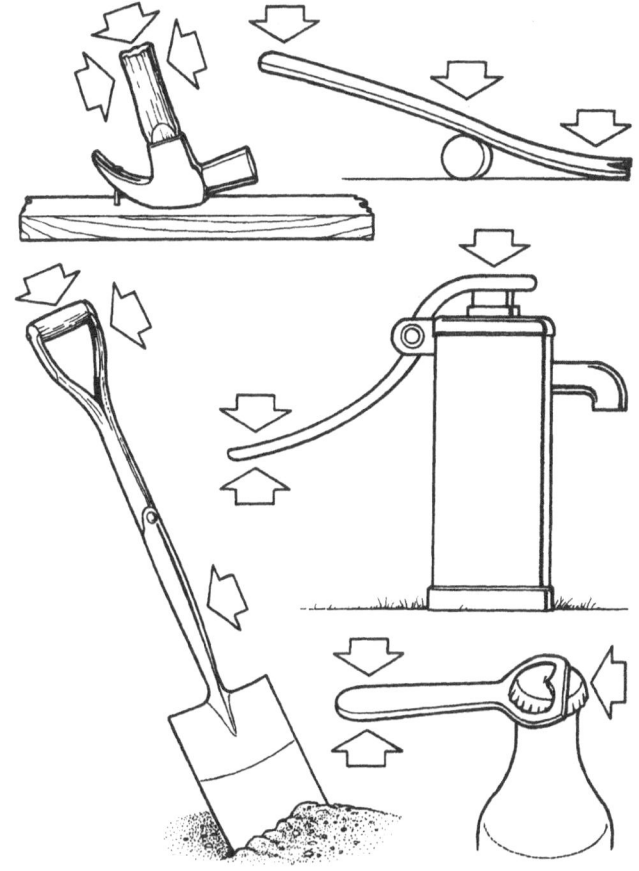

UNIT 27

Investigating bikes

You can learn a lot of science from your *bike*. For example, it uses levers in many ways.

A

Make a drawing of a bike, and colour in all those parts that act as levers.

B

Turn your bike upside down. Find out how many times the back wheel goes round for five complete turns of the pedals.

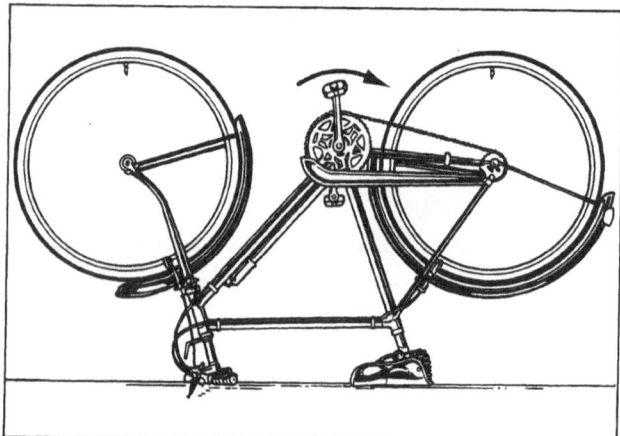

C

If your bike has gears, do the experiment in B again in *each* gear. Then fill in this table:

Gear	Number of turns of back wheel
First gear Second gear Third gear	

What conclusion can you draw from these results?

D

Kip lives where there are no machines such as tractors or electric motors. He wanted to use his bike to help to make work easier. This is what he did first:

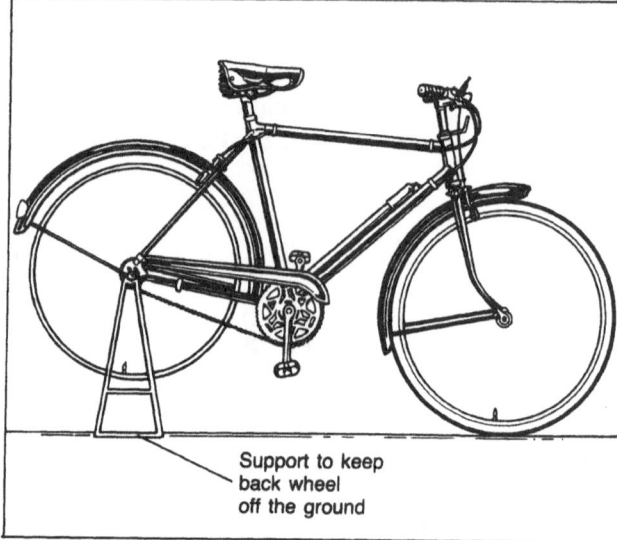

Support to keep back wheel off the ground

Then he took off the back tyre.
Explain how Kip could make his bike do useful work.

E

Touch the rubber of the brake blocks on your bike after you have been using them. You will notice that they are hot.

The heat is produced by *friction* (the rubbing of the block on the wheel-rim).

Look around your home or classroom and list other situations in which friction produces heat.

F

Bikes used to have solid tyres, but now they are *pneumatic* (have air in). The air is kept in by a *valve*.

You can make a valve with an old felt-tip pen. Take out all its parts and then make a small ball from foil which fits easily into the wide end of the plastic tube. It should be too big to come out of the narrow end.

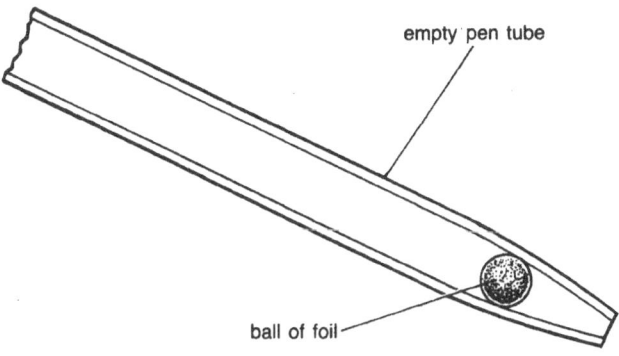

empty pen tube

ball of foil

Now draw a diagram to explain how your valve works.

G

Take the top off your bicycle bell, and watch how it works. Screw the top back, and find out how to get the best sound from it. Write down:
(a)   any questions you need to ask about it
(b)   all the things you have learned from examining your bell

H

Sairam works for a dairy in India. His job is to carry crates of milk cartons from the dairy to the shops on his bike like this:

This is very dangerous and the milk gets hot in the sun.

Can you design a milk carrier to fit on an ordinary bike? It has to be safe and it has to keep the milk cool.

I

Observe the different kinds of handlebars used on bikes. Write about the advantages and disadvantages of each kind.

UNIT 28

Investigating conservation

Conserving means *saving* or *protecting* things, for example, saving energy, saving *water*, protecting *forests*, protecting *wildlife*. Often we are wasting these things without knowing it. There is a danger that people all over the world will suffer one day because of this. So conservation is very important.

A

The chart below shows how much fuel people use on average in different parts of the world.

Country	Energy Used Per Person
Argentina	2,038
Bangladesh	41
Canada	13,453
China	835
France	4,995
Great Britain	5,637
India	242
Jamaica	1,390
Japan	4,260
Kenya	180
Nepal	14
Nigeria	83
Saudi Arabia	1,554
South Africa	3,479
Sweden	8,502
Singapore	6,211
Tanzania	53
Uganda	39
U.S.A.	12,350
U.S.S.R.	6,122
Vietnam	140
West Germany	6,627

(a) Where do people use the most fuel?
(b) Where do people use the least?
(c) What conclusions can you draw from the chart?

B

One of our main fuels is *oil*, which is used in cars, trucks, heating, plastics, margarine and many other things.

Make a list of ways in which you and your family could use fewer oil products without suffering.

C

This man is selling empty bottles and tins in a market because in his country people find them useful.

What do you do with empty containers?
Find out more about:
(a) bottle banks
(b) recycled paper
(c) what happens to your litter after it is collected

D

When you are in a supermarket, study how things are packaged and wrapped. Make a chart like this:

Product	Essential Wrapping	Non-essential Wrapping

Write down what you can learn from this.
Why do you think so much wrapping is used?

E

Here is a typical, large reservoir.

The forest surrounding it conserves water because tree roots help to hold water in the soil. Also forests encourage rain. If the forest was cut down, which of the following might happen?
(a)   The reservoir would dry up more quickly.
(b)   The reservoir would overflow.
(c)   The river would dry up more quickly.
(d)   The river level would be less constant.

F

This car runs on alcohol which is made from sugar-cane. Why is it better to use this fuel than petrol if possible?

G

The map of Kenya below shows that large areas of the country are preserved as wildlife parks.
People cannot live in the parks: you can only enter in a vehicle, and it costs about £2 each.

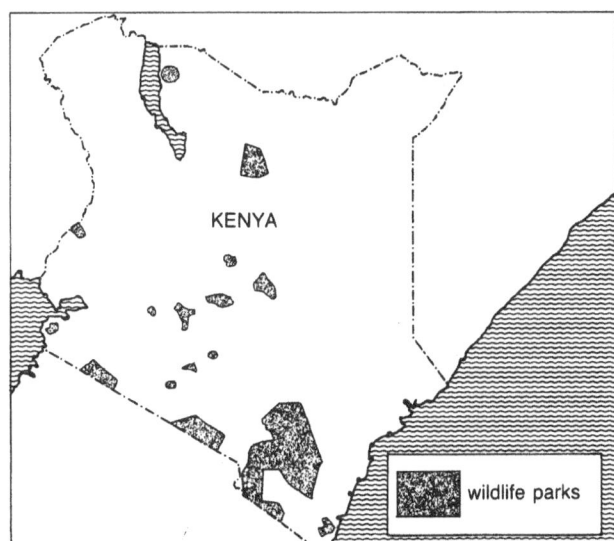

Discuss the advantages and drawbacks of having such parks in a country like Kenya, then fill in this table:

Advantages of parks	Disadvantages of parks

PART THREE: TESTS
TEST ONE

1 After a trip to the seaside, Mr Kelly's class asked a lot of questions about the things they collected. Pick out from the list all those questions which the class could answer by doing practical tests.
(a) How do pebbles become smooth?
(b) Which kinds of stone are hardest?
(c) What are the black bits in some pebbles?
(d) How do shells grow?
(e) Where did this rock come from?
(f) How much space is inside a shell?

2 Lorna knew that birds built their nests in different places, and she made this chart.

Nest site	Type of bird
Tree branches	Blackbird Crow
Tree trunks	Woodpecker
Buildings	Swallow Sparrow
On the ground	Gull Swan

She also made some notes about other birds. Where should she put them in her chart?
(a) Some bluetits made their home in a nestbox in our garden.
(b) Lapwings like to live on open moorland.
(c) Flocks of starlings live in the centres of cities.

3 Look at the picture below, and explain why it is not very clear.

4 Study this diagram of some apparatus.
(a) What is it for?
(b) How does it work?
(c) What is in the small dish?

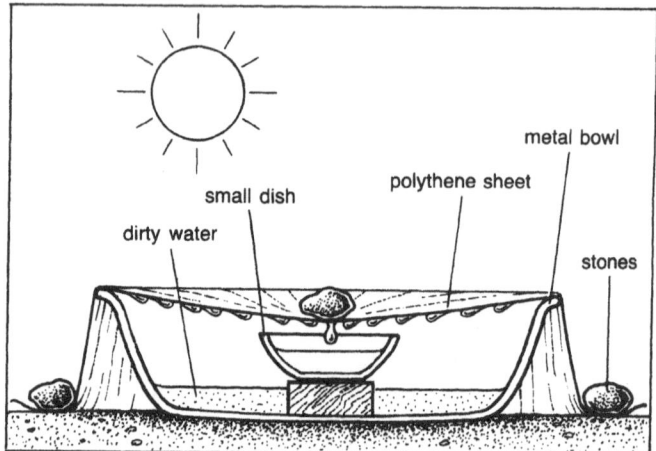

5 From the chart below, decide which day would be best for a day at the seaside, and explain why.

	Mon	Tue	Wed	Thur	Fri
Wind speed	10	15	20	35	30
Max. Temp.	16	15	19	14	14
Sun (hours)	3.1	5.8	8.0	11.0	0.2
Rainfall	0.4	0.0	0.0	0.1	0.3

6 John put two batteries into his calculator, and switched it on, but it didn't work. How should he complete this chart?

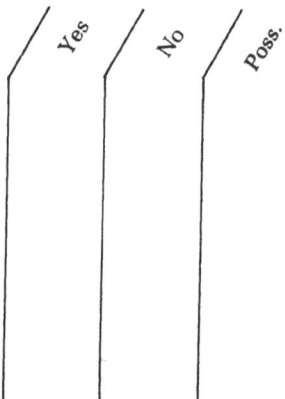

(a) The calculator is broken

(b) One battery may be dead

(c) Both batteries must be dead

(d) It needs more than two batteries

(e) Either the batteries or the calculator is faulty

7 All these structures use the same principle to make them work.

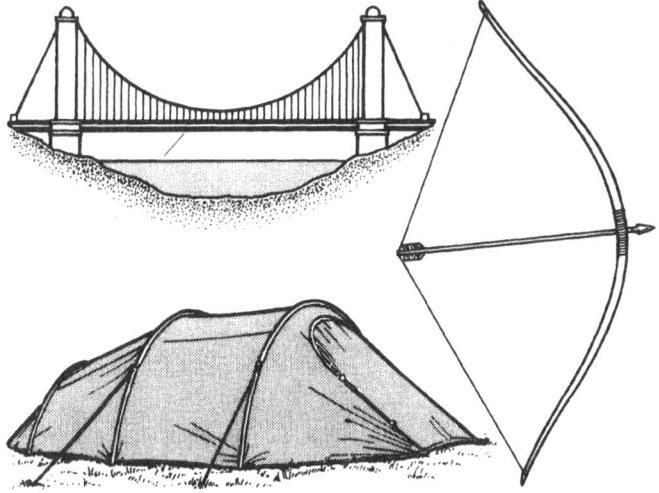

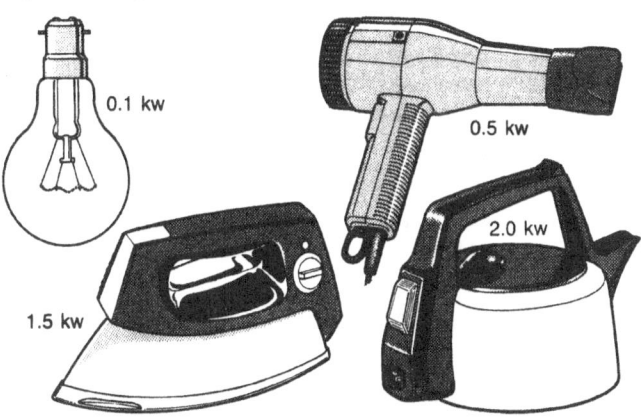

Name another which uses the same idea.

8 Here are four electrical appliances and the amount of power they use.

0.1 kw

0.5 kw

2.0 kw

1.5 kw

Complete this chart, and then write a suitable conclusion.

	Light	Heat	Move-ment	Power
Bulb				0.1
Hair-dryer				0.5
Iron				1.5
Kettle				2.0

9 Some predators prey on insects that are a pest on crops. If scientists bred these predators and released them on crops suffering from the pest, which of those would be the most important?
(a) Predators must be large in size.
(b) They must be easily seen.
(c) They must have a long life-cycle.
(d) They must not become pests themselves.

10 Look at the picture below.

(a) What is the woman doing?
(b) What problem will she have as the sun moves across the sky?
(c) How could this machine be modified to solve this problem?
(Answer using words or a diagram)

11 Marvin was trying to design a way to feed his goldfish when he was away on holiday. This was what he invented.

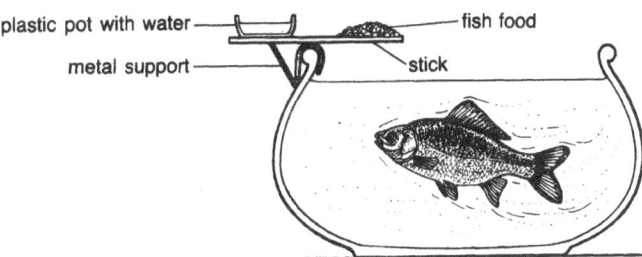

He explained that as the water in the cup evaporated, the balance would alter and the food would fall in the water.

However, Marvin didn't know how to make sure that the food would not fall until about the third day of his holiday. Suggest something he could do, to make it work as he wanted.

12 Read these instructions on how to make a blow torch.
1 You need a tin with a tight lid, a piece of cloth and a tube.
2 Make a hole in the lid for the wick (it should be quite large).
3 Make a hole for the tube in the side of the tin.
4 Make an air hole in the lid with a pin (it must be very small).
5 Put some paraffin in the tin.
6 Thread the cloth wick through the hole in the lid. Make sure the end dips into the paraffin.
7 Fit the blowtube tightly in the side.
8 Light the wick. Blow down the tube.
Now draw a diagram of someone using the blow torch, and label all the parts.

TEST TWO

1 Look at the drawings of the three fish below.

(a) List three ways in which they are all alike.
(b) Describe one thing about each fish which makes it different from the other two.

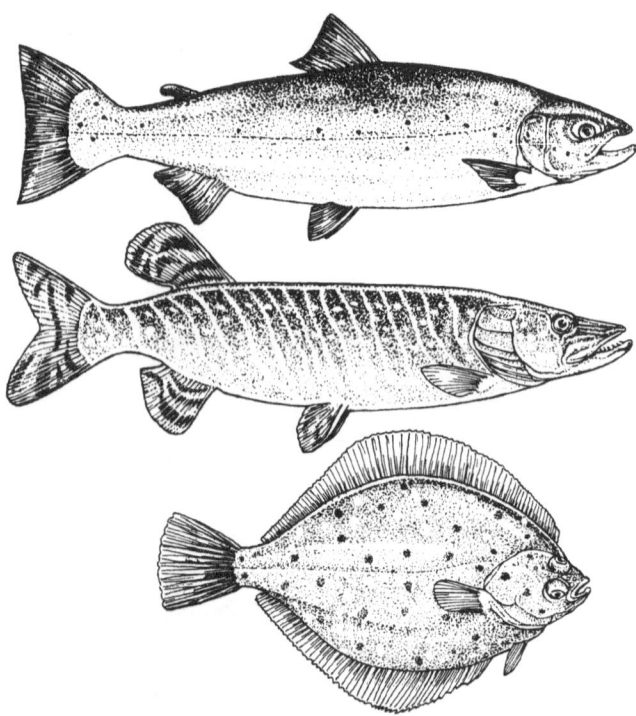

2 The toy wooden chicken below carries an egg balanced on its four wheels.

What happens to the egg when the chicken is pushed forward?
(a) It falls off
(b) It turns clockwise
(c) It wobbles one way then the other
(d) It turns anti-clockwise

3 Joel measured the height of a seedling each day for a fortnight. Here are his results.

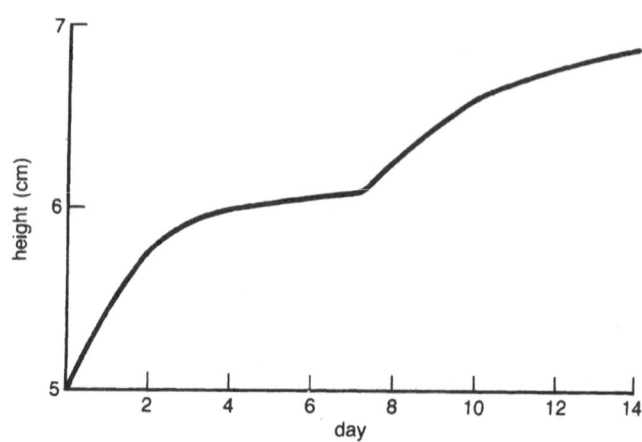

(a) How tall was the seedling when he first measured it?
(b) How much did it grow in the first four days?
(c) In which period did it grow fastest?
(d) What change happened after one week?
(e) What could have caused this change?

4 You have probably seen this sticker on the rear window of a car.

Draw two pictures to show what it looks like from inside a car:
(a) If you are sitting in the passenger seat looking at the rear window directly
(b) If you look at it through the driver's rear-view mirror

5 Cal used a large yoghourt pot to wash his hair in the bath. One day he was spinning it in the water: when it was empty, he noticed it would spin ten times without stopping, but when it was nearly full, it would spin only three times.
(a) How many times would it spin when it was half full?
(b) Write down three possible explanations for what Cal observed.

6 Gina poured four different liquids into a test tube. She shook the tube for ten seconds, then put it in a rack. After the tube had been standing for one minute, Gina drew what she saw. Here is her drawing.

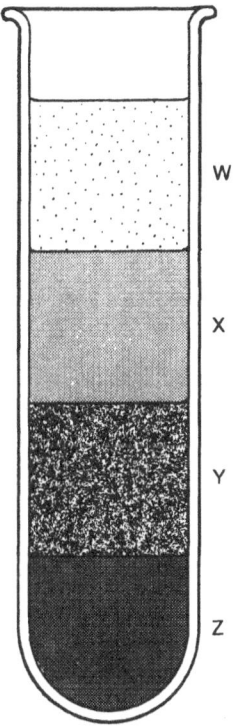

Which of these is the best conclusion to make from what you can *see* in the drawing?
(a) Liquid Y is denser ('heavier') than either W or X.
(b) The liquid Z is less dense ('lighter') than the others.
(c) The liquids are all different in colour.
(d) The four liquids will mix with each other when the tube is shaken.

7 Here are the lists of ingredients in three foods.
Food A: Sugar, Vegetable Oil, Hazelnuts, Cocoa Powder, Milk powder, Emulsifier, Flavouring.

Food B: Hydrolysed Protein, Wheatflour, Yeast Extract, Salt, Colour, Beef Stock, Vegetable Oil, Lactic Acid, Pepper, Onion Powder.

Food C: Wheatflour, Ryeflour, Sugar, Vegetable Oil, Dried Skimmed Milk, Salt.

(a) Which ingredient is in all three foods?
(b) Which has most Carbohydrate?
(c) Match each list to one of the foods in the list below:
    Oxo
    custard powder
    chocolate spread
    brown sauce
    crackers

8 Brian had two balloons full of gas. When he let them go, one floated upwards, whilst the other sank to the floor.

Which of these conclusions could *not* be true?
(a) One of the balloons was filled with air
(b) One was filled with a gas lighter than air
(c) Neither balloon was filled with air
(d) Both balloons were filled with hydrogen

9 Ee-Ling weighed a slice of bread, half a tomato, and a sausage. She put them all under the grill for five minutes, then weighed each one again after they had cooled.
(a) What question was she trying to answer?
(b) Show how she could best record her results.
(c) What do you think she found out from her experiment?

10 Ailsa wanted to find out if salt dissolves faster in warm water or cold. To do a fair test, which of these would she have to keep the same?
(a) size of jars used
(b) amount of water
(c) temperature of the hot water
(d) time of starting each experiment
(e) size of stirring spoon
(f) amount of salt added

11 Chris made a mobile from lollipop sticks, cotton and 2p coins, like this.

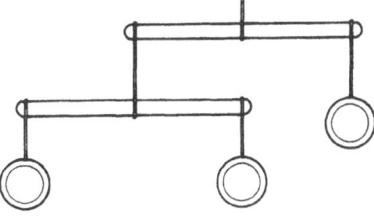

His mobile didn't balance, however, when he tried to hang it.
Draw a diagram to show how he can make it balance, without adding any more sticks or coins.

12 Look at this diagram of a minnow-trap, which is used to catch small fish.

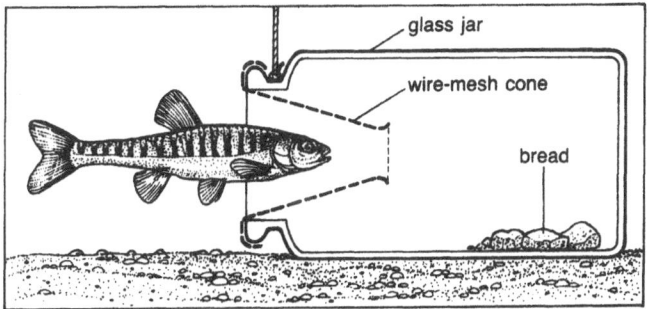

Using the same idea, draw a diagram to show how you could make a trap to catch *either* houseflies *or* garden worms.

PART FOUR: TEACHER'S SECTION

GRID PART 1 SKILLS TO TOPICS	Heat: temperature, conduction, expansion	Light: reflection, colour, lenses	Sound: music, speech	Electricity	Forces: weight, pressure, flotation, friction	Machines: tools, utensils, cars, levers/gears	Time	Weather: air, wind, rainfall	Surroundings: habitats, buildings	Materials	Liquids: water, changes of, state	Living Things: animals, plants, insects, bacteria	Growth/food: food, reproduction, human body	Structures: bridges, crystals, bones	Measuring
1 Asking questions				1A	1B	1B			1G	1A	1A	1C, 1D, 1E, 1F, 1G			1F
2 Observing	2H	2K, 2M, 2P, 14G	2Q, 5H, 14B			2B, 2C, 5F		2L, 7B	2A, 2E, 2F, 2J, 14C, 14F	2B, 2D, 2N, 14A		2G, 2M		2D, 2E, 2I, 3E	
3 Comparing	2H, 9A, 9E				9H	2C			2F	3D		1C, 3A, 3C	3B, 6D, 6E	3A, 3C, 3E	10I
4 Classifying	4D	5C				4B, 4D, 4E, 14E				4B, 4D, 4F, 14D		4C	4A, 4F		
5 Recording	6A	5C, 5G	5H	5D	10G	5D, 5F, 13E	5I		14K		9C, 11I	5A, 5B, 5E, 5H, 8B	6F, 7E, 6C	5E	5I
6 Interpreting	6A			6B	10G, 10H, 11K	7I, 12H		7B	6A	11K		5E	6C, 6D, 6E, 6F, 6G		7J, 10I
7 Analysing	7A					7D, 7E, 7I, 7J, 12F	7J	7B, 7C, 7D, 7H, 8G	7A	8G	7J		7G		10I
8 Concluding	8A	5G		8E				8G	8A, 8B	8G, 8H	8H	8B, 8C, 8D, 8F		8F	
9 Suggesting explanations	9A, 9E				9F, 9H				9J	8H, 9G	8H, 9A, 9C, 9G	9B, 9D, 9J	9B		
10 Predicting	10D				10A, 10E, 10G	10C, 10H	10C	10B		10F, 12A	10D, 10F, 10G, 10I, 15G, 15H	8C, 8D	12A		1F, 10H
11 Designing experiments	11C, 11E, 11H	15F			9F, 11D, 11F, 11J, 13G	7F	10C	7F		9G, 11C, 11G, 11K, 15C, 15T	11G, 11I, 13G, 15F	11A, 11B, 13G, 15B	9B		11F, 11J
12 Applying ideas	12B, 12E, 12G, 12I	12F, 12I	12D	12I	12C	12B, 12C, 12D, 12E, 12F, 12G, 12H			12A	3D, 12A, 12B	12A, 12E, 12I	12I		12F	
13 Improvising		13A		5D	13F, 13G	7D, 7J, 13A, 13B, 13C, 13F	7J	7D		13F	13G	13E, 13G			13D
14 Being curious		14F, 14G	14B	1A		14E			14C, 14H, 14K	14A, 14D		1G, 14D		14J	
15 Being practical	12I, 15E	12I, 15E, 15F, 15I		12I	15D	1B, 13A	15D	15A		15C, 15J	15D, 15F, 15G, 15H	11B, 12I, 15B	9B		

GRID PART 2 TOPICS TO SKILLS	Asking questions	Observing	Comparing	Classifying	Recording	Interpreting	Analysing	Drawing conclusions	Suggesting explanations	Predicting	Designing experiments	Applying ideas	Improvising
16 Earth		16A, 16B, 16E			16B, 16E, 16F	16A, 16D	16C, 16D	16D, 16B	16E			16F	16B
17 Air	17A, 17B, 17I	17A, 17B, 17C, 17D, 17E, 17F	17G			17D, 17F	17F	16D, 17E, 17F, 17G	17B, 17C		17H	17F	17H
18 Water/liquids		17F, 18B, 18H	18B	18A	18B, 18E	18D, 18F	18D, 18F	18D, 28E	18C	18G, 28E	18E	17F, 18C	18E
19 Weather	17I, 19E	19D		19D		19A, 19F, 19G	19C, 19G	19F, 19G	18C	19A, 19B		19A, 19F	
20 Natural habitats	20C, 20D	16B, 20A, 20B	20B, 28G	20H	20F, 20G		20E	20E	20E	28E	20D, 20F, 20G		20G
21 Man-made habitats	21C, 21H	21A, 21D, 21G, 21H, 24B, 26B, 27E	21A, 21E, 21H	21H, 26B	21A, 21C, 21D, 21E, 21F, 24B	21D, 21G, 21H	21F, 27E	21A, 21F	21B, 25D	21I	21F	21E, 21I, 27E	
22 Reproduction	22B, 22F	22D, 22E		22D, 22E	22D	22E	22E		22A		22C		22C
23 Food	23E	23D	23F	23B	23B	23A, 23C	23A, 23F	23A, 23F	23E	23C	23G	23C, 23E, 23G	23G
24 Health	23E, 24A, 24D, 24G, 24H	23D, 24B, 24D	17H	24F	24A, 24B, 24C, 24D	24E	23D, 24B	24A, 24H	18C, 23E, 24E	23C, 24C		23C, 24B, 24C, 24H	24E
25 Work	25G		25A, 25C, 25G	25E	25D, 25G, 26E	26G	25B, 25C, 25F	25G	25C, 25D	25C	26E	25G, 26D	26E
26 Tools	26D	26B, 26F, 27G	21E, 26A	25E, 26A, 26B, 27A	26E	26C, 26G	26A, 26B, 26F, 26G	26C, 26G	26B	26C, 26F	26E	26C, 26D, 26G	26E
27 Bikes	27G	27A, 27B, 27C, 27E, 27G	27I	27A	27A, 27C, 27F		27A, 27C	27C, 27I	27F	27D	27F, 27H	27D, 27G, 27H	27D, 27F, 27H
28 Conservation	20C, 28A, 28C		28G	28D	20C, 28D	28A, 28E	28A, 28B	28A, 28D, 28G	28D, 28F	28E		23E, 28B, 28F	

TEST GRID

	TEST ONE	TEST TWO
Asking questions	1.1	2.9
Observing	1.3, 1.4, 1.7, 1.10	2.4
Comparing		2.1, 2.7
Classifying	1.2, 1.7	2.7
Recording	1.2, 1.6, 1.8, 1.10, 1.12	2.6, 2.9, 2.12
Interpreting	1.4, 1.5, 1.7, 1.8, 1.12	2.3, 2.6
Analysing	1.3, 1.5, 1.7, 1.9	2.3, 2.4, 2.6, 2.7, 2.11
Drawing conclusions	1.2, 1.5, 1.6, 1.8	2.3, 2.6, 2.8, 2.9
Suggesting explanations	1.1, 1.3, 1.5, 1.6	2.3, 2.5
Predicting	1.10	2.2, 2.4, 2.5
Designing experiments	1.11	2.10
Applying ideas	1.7, 1.11	2.11, 2.12
Improvising	1.4, 1.10, 1.12	
Being curious	1.1	2.9, 2.10
Being practical	1.1, 1.11	2.12

ANSWERS TO QUESTIONS IN UNITS

The list of answers below is incomplete, because
(a) answers to worked examples are given in the text itself
(b) some questions are open-ended, with many possible answers
(c) some answers will depend on local surroundings and circumstances.
Hence this list gives only those answers which apply generally to all circumstances.

Unit 1 Asking questions

A The water bends towards the comb, but not always by the same amount, depending on the type of hair, material of the comb, humidity etc.
F Breathing rate and pulse rate increase with exercise.

Unit 2 Observing

D Crystals are lozenge-shaped.
G 1 Rump 2 Face 3 Back 4 Legs
I 1 12 edges 2 6 sides 3 cube-shaped
K Left hand
P The lamp post appears to move backwards.
Q The pitch of the sound goes lower as it passes.

Unit 3 Looking for similarities and differences

D One is made from reeds, the other from plastic. One is woven, the other is moulded. You might compare size, strength, flexibility, price, etc.
E 1 Bailey Bridge—rigid girder (box)
   2 Suspension Bridge—road hangs by steel cords from towers
   3 Pre-stressed concrete sections—stand on piers

Unit 4 Classifying

A Fruit/Vegetables grouping is more useful than colour.
B Group 2
D Conductors are saucepan, spoon, kettle.
E Bottle opener—lever. Key—lever.Corkscrew—cutter and lever.

Unit 5 Recording

A 1 nest and young—drawing
   2 birds comings and goings—time chart
   3 behaviour of young birds—written description
   4 food brought by parents—list
D The torch is made from a beer can, sellotape, rubber band and metal strip. A hole is made in the base of the can for the bulb, and the batteries are fixed at the other end by sellotape. The metal strip is held in place by the rubber band, and the torch is switched on by pressing the strips of metal together.

Unit 6 Interpreting

D 1 4 kilos
   2 6 kilos
   3 It went down.
   4 About 4 months old
E Statement 2 is false.
F 1 5 feeds
   2 3 spoonfuls
   3 1,000 millilitres per day
G 1 Measure boiled water into bottle.
   2 Carefully measure level spoonfuls of milk powder.
   3 Add powder to bottle.
   4 Put the cap on and shake the bottle to dissolve the powder.

Unit 7 Analysing

A Yes
B The north-east and Scottish coasts will have fine weather. The west coast should be cloudy.
D As the wind blows, it turns the needles/vanes on the top. The faster the wind blows, the faster it turns.
E It is top heavy; the needle through the bottle may not turn easily; it needs fixing at the base.
F Put the vertical needle inside a hollow tube (e.g. an empty 'BIC' pen), then fill the bottle half-full with sand, to stabilise it.

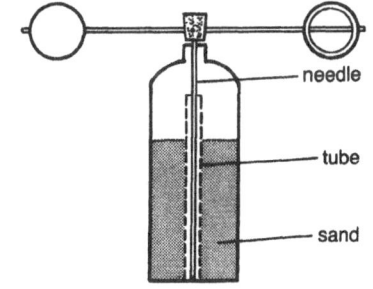

G Yes
H No, because he never caught fish on warm, sunny afternoons.
I 2 turns anti-clockwise, 3 turns clockwise, 4 turns clockwise.
J 1 No, because the water runs out quickly at first, but only slowly when the bottle is nearly empty.
   2 1.15

Unit 8 Concluding

D 1 Birds, fish
   2 Fish would eat more flies, frogs would eat more beetles. Less grass would be eaten.
E The bulb only lights when the wires are connected to the opposite ends of the battery and the side and base of the bulb, to make a circuit.
F A and D are insects. Insects have 6 legs.
G (1)–(b) (2)–(d) (3)–(a) (4)–(c) (5)–(e)
H 1 Liquids behave differently when put on different surfaces; for instance, water soaks into newspaper, but not the others.
   2 Because oil and water cannot soak into it or pass through it.

Unit 9 Suggesting explanations

E The cups were different sizes; they were made of different materials; the amount of tea was different in each cup; the cups were left in different places to cool.
F The nails were dropped from different heights; the bottle-top was not evenly loaded; the water was not calm on the surface.
G All the following things could have varied: amount of soil, amount of water, size of cotton wool plug, size of hole in funnel, composition of soil, dryness of soil, speed of pouring water.
H Sliding could be affected by all except 1 and 2.
I Because hot water rises and cold water sinks, so they stay separate (until the hot water cools).

Unit 10 Predicting

E Four eggcupfuls
F 2
G It will fall between the other two (answer to be represented in a diagram).

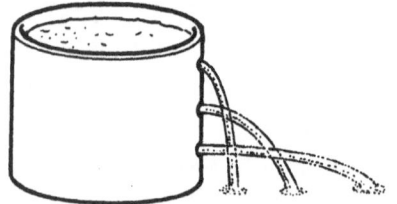

H Rover: 16.2 Ford Escort: 43.5 Vauxhall Cavalier: 29.7
I More water will spill over.

Unit 11 Making tests fair

D Drop nails from same height. Keep water calm. Keep the bottle top evenly loaded.
E Keith's test is not a fair one, because the candles give different heat, and the rods are of different lengths and thicknesses.
F 2,3,4
G All except the type of soil
H 1,2,4
I (1) same (2) same (3) two (4) both
K same: type of match, type of clip different: type of glue, amount of glue

Unit 12 Applying ideas

D Guitar, violin, piano, etc
E Thermometer
F Parabolic reflector
G Glider
H

	Slowly	Quickly
Bicycle	√	√
Car	√	√
Clock	√	
Whisk		√

I 1 photographic printing
2 steam engine
3 dressing to keep warm/cool
4 preserving food by drying or sealing
5 insulators (e.g. screwdriver handles)

Unit 13 Improvising

C 1,3
D Stick a strip of paper up the outside of the bottle. Add 5ml of water at a time from the spoon, and mark the level on the paper each time, until the bottle is full.
F The turbine should be made of foil. Very little water should be used. The hole should be as small as possible.

Unit 14 Being curious

E Things which open and close.
J apple, logs, pipes and telephone

Unit 15 Being practical

C 2
G Different colour patterns appear.

Unit 16 Investigating earth

A (a) Garden being dug over with a fork; peat turves being cut and stacked; iron ore being excavated and loaded.
(b) As a growing medium for plant foods; as a source of fuel; as a source of essential minerals.
D Soil B

Unit 17 Investigating air

A By watching things blown by the wind, divers under water, etc.
B Air pressure inside the balloon forces it to fly through the air.
D (a), (d)
E Misty condensation of tiny water droplets. Water vapour.
F (a) bubbles adhering to sides of glass.
(b) they were previously dissolved in the water.
(c) dissolved air is essential to the life of fish and other water life.
G (a), (d)

Unit 18 Investigating water and other liquids

A Oil, milk, vinegar, lava, tea.
C The large stones at the top remove bulky impurities like leaves, dead creatures, first.
D Puddle water was dirtiest, tap water was cleanest.
F The temperature remained at 0°C until all the ice had melted, then it slowly began to rise.
G Throat goes dry when mouth is kept open; felt tip pens don't work when left with the top off; pavements dry after rain stops.

Unit 19 Investigating weather

A (a) cold and wet
(b) cold and dry
(c) warm and dry.
B It is likely to get wetter.
C A is under a roof, C is under a tree, D is too near a wall (splashes). So only B will give a fair reading, because it is in the open playground.
D Heavy rain
Fine and sunny (Cumulus)
Fine but changeable
Thundery (Cumulonimbus)
F (a) November
(b) July
(c) Nairobi
(d) March

G (a) Riyadh
(b) Corfu, Munich
(c) Edinburgh
(d) Early summer, or perhaps autumn

Unit 20 Investigating natural habitats

D Which creature lays them? How long do they take to hatch? Does any creature eat them? What damage is done to the leaves?
E (a) deprives flies of breeding grounds and food—so less flies
(b) deprives snails of food—so less snails
(c) increases number of flies and snails
(d) numbers of fish increase
H Monkey, koala, woodpecker, bluetit, caterpillar: tree dwellers
Seal, penguin, gull, crab: sea creatures
Rabbit, fox, mole, kingfisher, badger: live in holes in ground

Unit 21 Investigating the man-made environment

B For security, source of food, warmth

Unit 22 Investigating reproduction

A Most of the salmon eggs will be eaten or destroyed. Few will survive, so many are needed to ensure the continuation of the species.
B Bees carry pollen on their legs from one flower to another, and pollen fertilises flowers.
C Count the number of eggs hatched each day from birth onwards. Make a table. Find which day has most hatches.
F (a) The period of time immediately before birth, when the baby is moving out of the uterus. Can be up to 24 hours.
(b) muscle movements which help push the baby out.
(c) giving birth without any drugs, surgery or specialist treatment.
(d) a cord which attaches mother to baby. It is cut after birth.
(e) attached to side of womb and nourishes baby. Comes out after baby is born.
(f) someone who is trained to help a mother during childbirth.
(g) baby is born upside down (head comes out last, instead of first).

Unit 23 Investigating food

A (a) true (b) true (c) false
(d) true (e) true

B Bodybuilding and regulating: liver, eggs, bread
Bodybuilding and energy: cheese
Energy and regulating: raisins
Regulating only: oranges, carrots, apples
All three: baked beans

C About 850g

E Helping them to grow their own food and be self-sufficient, by such means as providing land, loans to buy land/machinery, irrigation, better seeds etc. Also by a change in the way prices of exported foods are controlled.

F Cooking—meat   Pasteurising—milk
Drying—grapes   Freezing—fish
Canning—peaches   Pickling—onions

Unit 24 Investigating health

C (a) Use a handkerchief. Avoid coughing over people.
(b) Wash your hands. Use toilets carefully.
(c) Wash clothes and hair often. Avoid too much close contact with others.

E (a) To prevent contact with animals, children, and to allow safe seepage.
(b) To keep out flies.
(c) To prevent contamination of water and food supplies.

F Need a doctor: deep cuts (tetanus), mumps, pneumonia, earache.
Treat at home: indigestion, boils, stings, sprains, chickenpox.

Unit 25 Investigating work

A (a) running   (b) swimming
(c) digging   (d) climbing stairs

B (a) muscle, petrol   (b) steam, electricity, diesel   (c) wind, petrol, steam   (d) sun, electricity
(e) wind, electricity, water

(f) wind, petrol, muscle

C (a) on head, in barrow, on back
(b) in each hand
(c) it needs less use of muscles when the weight is distributed directly over the spine.

E Set A: (a), (d), (e), (h)
Set B: (b), (i)
Set C: (c), (f), (g)

F (a) saw, axe, power saw   (b) plane, sandpaper   (c) saw, chisel, vice, nail, screw   (d) spade   (e) ladder, secateurs, basket

Unit 26 Investigating tools

A Penknife—cutters. Tap—turners. Nutcracker—openers.

C Long handled screwdriver with narrow handle

E A lever could be made from a pole or strong stick, with a small stone or piece of wood as a fulcrum.

F (a) one on right goes up (b) they just balance (c) at equal distances from the centre (fulcrum).

G

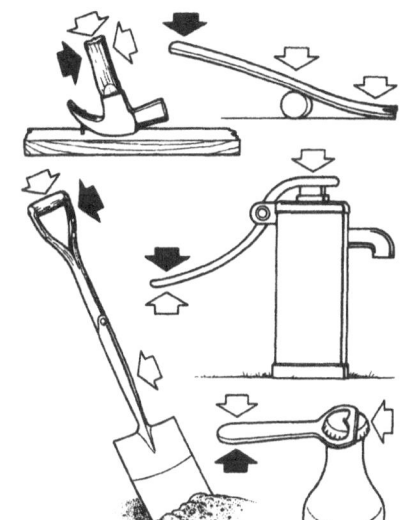

Unit 27 Investigating bikes

D It could be used to turn a pump, generator, lathe etc, by attaching some kind of drive to the back wheel.

F Line a box with polystyrene or some other insulator to keep the contents cool. Design a rickshaw or trailer to make it safe to carry.

Unit 28 Investigating conservation

A (a) Canada, USA, Sweden, USSR, West Germany—western 'developed' countries
(b) Nepal, Bangladesh, Uganda, Tanzania—'developing' countries of Asia, Africa
(c) Inequality of use is unfair. Western countries are exhausting the world's reserves of fossil fuels. Developing countries cannot afford to use modern technology.

B Insulate house; drive smaller car; use less packaged/processed food.

E (a), (d)

F Alcohol from sugar cane is renewable, petrol from oil is not.

G Advantages: preserves wildlife, brings income from tourists.
Disadvantages: uses up valuable land, disturbs natural habitats, prevents most people from seeing animals (no cars).